For Sylvia — thanks for reading it.
And for Henry, James and Rupert — you can read it later.

Dynamics of Biological Systems

Michael Small

CRC Press
Taylor & Francis Group
Boca Raton London New York

CRC Press is an imprint of the
Taylor & Francis Group, an **informa** business

A CHAPMAN & HALL BOOK

CRC Press
Taylor & Francis Group
6000 Broken Sound Parkway NW, Suite 300
Boca Raton, FL 33487-2742

© 2019 by Taylor & Francis Group, LLC
CRC Press is an imprint of Taylor & Francis Group, an Informa business

No claim to original U.S. Government works

ISBN-13: 9781439853368

Visit the Taylor & Francis Web site at
http://www.taylorandfrancis.com

and the CRC Press Web site at
http://www.crcpress.com

Contents

Preface

Biological systems exhibit rich dynamic behaviour over a vast range of time and space scales: from the spontaneous rapid firing of cortical neurons to the spatial diffusion of disease epidemics and evolutionary speciation. This text unifies many of these diverse phenomena and provides the computational and mathematical platform to understand the underlying processes. Through an extensive tour of various biological systems, we introduce computational methods for simulating spatial diffusion processes in excitable media (such as the human heart), and mathematical tools for dealing with systems of nonlinear ordinary and partial differential equations (necessary to describe neuronal activation and disease diffusion). We show that even relatively simple mathematical descriptions are capable of capturing a wide range of observed biological behaviour and offering insight into the dynamic system hidden within. We present mathematical models and computer simulations that can provide insight into cardiac pacemakers and artificial electrical defibrillation, and can suggest appropriate control strategies to mediate the effects of past and future pandemics.

The key concept of this text is the idea of the model: a computational or mathematical abstraction of a more concrete description. At various layers, the models we use to describe specific biological systems become more abstract and hence more manageable. At the same time, the scope of the model becomes broader. The same basic mathematical description can describe very different biological phenomena. There is a natural convergence between models of hormone secretion, respiratory carbon dioxide uptake and ecological population fluctuations. Starting from each separate biological system, we can obtain models that at various levels of abstraction are equivalent. Consequently, when the models exhibit gradual loss of stability and onset of chaotic dynamics through a sequence of period doublings, we can begin to understand the origin of the same characteristic pattern in each of the diverse biological systems.

Via a range of models, we observe a host of different bifurcations and dynamic structures in a wide range of both temporal and spatial systems. Consideration is also given to emerging research areas in complex biological systems: pattern formation and flocking behaviour, interaction of autonomous agents, hierarchical and structured network topologies in a range of systems (from neuronal aggregations in the nematode worm to synchronous behaviour in herd animals and disease transmission dynamics). Tools from complex systems and complex networks are introduced and demonstrated to be useful for dealing with a range of real phenomenological systems. Biological systems exhibit a rich assortment of coherent and emergent behaviour. Biological systems fluctuate in time and their behaviour changes over time. This text introduces the necessary mathematical and computational tools to understand these temporal phenomena.

The target audience for this text is undergraduate students with a basic mathematical foundation. High-school level or first-year undergraduate calculus should suffice: substantial knowledge or proficiency in differential equations is not required. Some knowledge of computing programming is helpful, but not necessary. The material in this text has been used for several years as the basis of a one-semester bioengineering course taught to electronic engineering undergraduates. One of the main aims of this course has been to teach a broad range of mathematical and computational modelling skills and show students the real practical application of these skills. In addition to undergraduate students with some calculus background, this text will be of interest to researchers in biological and physiological sciences who need an overview of contemporary mathematical and computational techniques. Beginning postgraduate students in chaos, complexity and nonlinear dynamics will benefit from the broad range of techniques and contemporary tools that are discussed.

Exercises for the reader, as well as computational and research projects, are suggested at the end of each chapter. The website for this book will provide additional resources including MATLAB® code and example data sets.

Acknowledgements

During the writing of this book I was provided with financial support from the Hong Kong Polytechnic University (Internal grant G-YG35 and G-U867) and funding from the General Research Fund of the Hong Kong University Grants Council (PolyU 5279/08E and PolyU 5300/09E).

MATLAB® is a registered trademark of The MathWorks, Inc. For product information, please contact:

> The MathWorks, Inc.
> 3 Apple Hill Drive
> Natick, MA 01760-2098 USA
> Tel: 508-647-7000
> Fax: 508-647-7001
> E-mail: info@mathworks.com
> Web: www.mathworks.com

Finally, the cover illustration is the work of Henry Small (aged 7).

Feedback

Although the best way to really understand how these techniques work is to create your own implementation of the necessary computer code, this is not always possible. Therefore, I have made MATLAB® implementations of many of the key algorithms and the various case studies, available from my website (http://small.eie.polyu.edu.hk/). You are most welcome to send me your comments and feedback (small@ieee.org).

Michael Small
Hong Kong

Chapter 1

Biological Systems and Dynamics

1.1 In the Beginning

Biological systems change over time; they change over time in two important and very different ways. First, animals breathe, populations rise and fall and your heart keeps beating. All these changes are regular and continual. In each case the output of the system is changing in a rhythmic manner and past behaviour provides (at least some) indication of what will happen in the future. This change is the continuous fluctuation that makes up the rhythm of life, and varies within some acceptable range. The system is in some sort of balanced equilibrium. Mathematically, the system is said to be *stationary*, not because the system is not moving, but because the mathematical description of the system and the system parameters are not changing.

There is a second type of change in biological systems. This occurs when the system itself (and by extension, the mathematical description of that system) changes. For millennia the global climate system has been in a quite stationary state — certainly there is seasonal variation and longer climatic cycles , but all this variation is within some natural bounds. Both sorts of change are illustrated in Fig. 1.1. Relative global warm and cool periods occur and fluctuation occurs, but the system is stationary. Then, man dramatically increases the carbon dioxide content in the atmosphere to a level which is unlike anything in natural history and at a speed far faster than nature has ever managed. This man-made change in the state of one variable in the global climate system means that the system is now nonstationary. Its behaviour is entering a regime unlike any that has been seen in the past.

Fig. 1.1 illustrates the change in a system, the human cardiovascular system, as a patient suffers a heart attack. Similarly, individual neurons within your brain are rapidly and spontaneously pulsing. The neurons (and there are about one thousand billion of them) in your brain are the individual cells which process information and are collectively responsible for consciousness and thought. Despite this massive responsibility, the state of each individual neuron can be fairly well described with a simple set of mathematical equations. The behaviour of each neuron can be qualitatively predicted with reasonable accuracy — by *qualitative* prediction we mean that we are able to provide a description of the type of predicted behaviour, without predicting

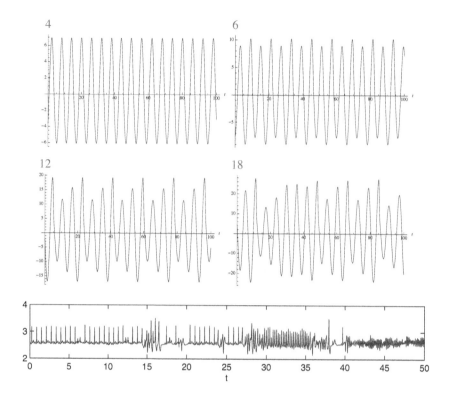

Figure 1.1: Dynamics of systems. The top four plots show the stable dynamic behaviour for the theoretical Rössler system ($\frac{dx}{dt} = -y - z$, $\frac{dy}{dt} = x + 0.1y$, and $\frac{dz}{dt} = 0.1 + z(x - c)$). The horizontal axis is time and in each case the system changes continuously. For a parameter (c) value of 4, the system is periodic. However, as we change the value of the parameter, the behaviour of the system changes. For $c = 6$ the system is now bi-periodic and for $c = 12$ the system becomes period three. Of course, for all intermediate values of c the system behaviour changes gradually. But for any fixed value of c, the system will oscillate — changing continuously. Finally, for $c = 18$, the system behaviour is bounded and not periodic — it exhibits what is mathematically defined as *chaos*. The lower panel illustrates a real recording of the electrocardiogram of a human and shows a second example of the change in dynamics between (in this case, at least) three states: from a stable (almost periodic) regular heartbeat on the left, to ventricular tachycardia in the middle and ventricular fibrillation on the right. This trace illustrates the progression of onset of a heart attack and in this case the patient (in a coronary care unit) only recovered after medical intervention. Within each window, past behaviour provides (some) guide to the future. But when the system parameter changes, this is not possible because it is not possible to predict how the system parameter changes.

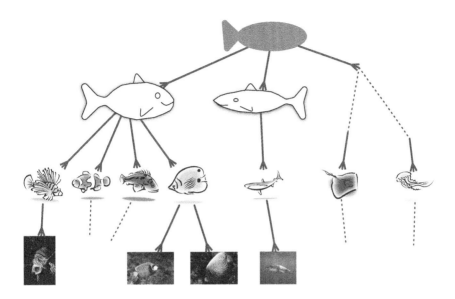

Figure 1.2: **The hierarchy of models.** Specific highly detailed models of individual species of fish are shown along the bottom. At higher levels, these are then abstracted to cartoons of general types of fish — each encompassing multiple species and varieties from the lower level. These cartoons can be further abstracted to general fish body morphologies (in this illustration one could also consider the analogous zoological groupings); these morphologies are then abstracted further into a vague notion of fish shape. At each level the illustrations take the role of a model — describing certain features of the things being modeled, while ignoring others. More general models become simpler as they encompass more varieties — but account for the varieties with less and less detail. (Note that for clarity, some portions of this diagram are omitted and shown as dotted lines.)

the precise values of the system (we will see later why such precise numerical predictions, *quantitative* predictions, cannot be made). However, as your attention shifts from this text to the activities of last Friday night (whatever they were), delicate chemical signals in your brain change the balance between different neurotransmitters (the family of chemicals responsible for the transmission of information between different brain regions, and between individual neurons) within different regions of the brain and this causes the activity of these individual neurons to change. The parameters of the mathematical description of each neuron needs to change, and hence the qualitative behaviour changes.

This book is about both sorts of change: the way in which a biological system in a stable equilibrium changes, and the way in which systems can be perturbed from equilibrium and then settle to a new equilibrium. The main tool we will employ to describe and to study this change will be *models*. The models may be mathematical, computational or even biological (in the case where one systems acts as an archetypal representative for another). However, models always incur a cost. Each model we use is a simplification and abstraction from the real system on which we wish to focus. The models are therefore imperfect. But this imperfection has two great benefits.

First, if we choose the right models, then the model contains all the important behaviour of the original system, but at the same time it is somehow simpler (see Fig. 1.2).

Second, and more dramatic, by performing the right simplifications we will find the same model description in many different systems. We can see the same principle that governs population dynamics can also be used to describe hormone secretion and regulation of respiration. Hence, a good description of the dynamical behaviour of the model, and of how that model's behaviour can change, can be applied to a host of different biological systems.

This book is about models and about how the changing dynamical behaviour of models can be described. We will begin this introduction with two examples to illustrate the main principles: the hemodynamic system and Cheyne–Stokes respiration.

1.2   The Hemodynamic System

The human body contains between five and six litres of blood distributed in a closed system over a circulatory system consisting of around 100,000 km of pipes. That is, from the largest blood vessels delivering blood to and from the heart to the most distant and minute capillaries. Obviously, with these dimensions, most of the tubing that makes up the circulatory system is very very tiny. The smallest capillaries are 8 μm in diameter and have walls that are no more than one cell thick. The reason for this minute scale and vast length is that, as a consequence, the entire human body is densely packed with blood vessels. It is estimated that every cell in the body is no more than 0.1 mm from a capillary. Because the capillaries themselves are so thin, and their walls consist of a layer of single cells, it is easy for carbon dioxide and oxygen to diffuse through them.

The oxygenated blood is delivered from the heart to the body (refer to Fig. 1.3). The oxygen is provided to the muscle tissue and other organs in the body and replaced in the blood by carbon dioxide which is pumped back through the heart to the lungs. Of course, once in the lungs, the carbon dioxide is dumped from the blood and replaced by oxygen to repeat the entire cycle. Hence the circulatory system consists of a figure-eight pattern with the heart

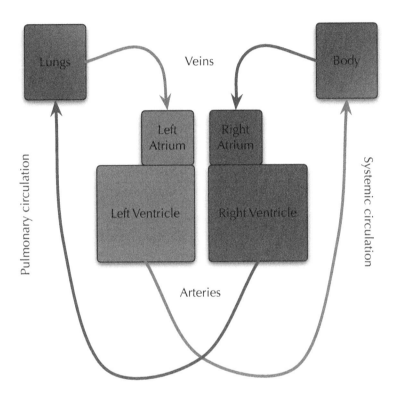

Figure 1.3: **A model of the human circulatory system.** The four chambers of the heart are represented in the centre and shown to drive oxygenated blood to the body and then back to the heart before passing the now-deoxygenated blood to the lungs where the blood is replenished with oxygen and returned to the heart.

at the centre. Both oxygenated blood from the lungs on its way to the body and deoxygenated blood from the body returning to the lungs pass through the heart. The deoxygenated blood passes through the right chambers of the heart on the return trip to the lungs and the fresh oxygenated blood passes through the left chambers. On each side there are two chambers: an *atrium* and a *ventricle*. The left atrium receives the blood from the lungs and passes it to the left ventricle and then onto the body. The right atrium receives blood from the body and passes it to the right ventricle before returning to the lungs. The circulatory system is divided into two sections, one side of the figure eight, dealing with blood flow to and from the lungs is the *pulmonary* circulatory system. The other side of the figure eight is connected to the body and is known as the *systemic* circulatory system.

The heart acts as a pump and essentially drives the blood through the body. Hence, the blood leaving the heart through blood vessels known as *arteries* is at high pressure, while the blood returning to the heart through *veins* is at low pressure. Within the heart itself, the atria are relatively small upper chambers of the heart and connected via a valve to the much larger lower chambers. It is these lower chambers, the ventricles (and in particular the left ventricle) that do most of the work of the heart. In a healthy human at rest, the left ventricle pumps about 72 beats/minute and about 70 mL/beat. In doing so, the left ventricle generates about 1.7 W of mechanical power. During vigorous exercise, the body's demand for oxygen increases and the heart rate can increase to over 160 beats/minute.
indexcirculatory system!heart

Figure 1.3 depicts a simple model of the human circulatory system with the various nomenclature described previously. In Fig. 1.4 we simplify that model to another model in which the circulatory system is represented as a pipe in a figure-eight configuration with a pump at the central crossing point. Now consider a section of that pipe. Or, consider a major artery leaving the heart. How is the physical force exerted by the heart to pump blood related to the physical flow realised in the pipe?

We start our description of this system with another model — represented in Fig. 1.5. Here we focus on a single cylindrical section of pipe and we apply the standard mathematical model of fluid flow to describe the movement of blood in that pipe. The Navier–Stokes equation[1] is a second-order partial differential equation describing the spatial-temporal relationship between the flow of a fluid $u(x, y, z, t)$, pressure P, fluid density ρ and viscosity ν. Note that the function u describes the instantaneous velocity as a function of time and space, and this function satisfies the following differential equation,

$$\frac{\partial u}{\partial t} + u \cdot \nabla u = -\frac{\nabla P}{\rho} + \nu \nabla^2 u. \tag{1.1}$$

---

[1] The Stokes of Navier–Stokes is not the same as the Stokes of Cheyne–Stokes.

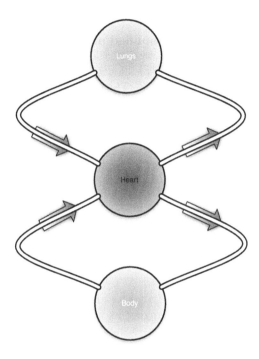

Figure 1.4: A simplified model. The human circulatory system of Fig. 1.3 is represented here as a more simplified model. In this structure, the heart is a single pump driving blood flow between the heart and the body. From this diagram we now simplify things further and focus only on the driving force and the flow of blood in a single section of the circulatory system — see Fig. 1.5

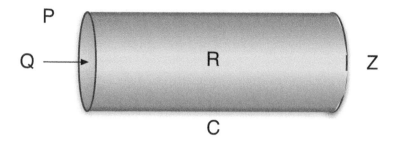

Figure 1.5: **An even more simplified model.** Now we consider only one part of the human circulatory system — the flow of blood in a cylindrical, permeable and elastic walled pipe. The flow Q is driven by the forcing pressure P, subject to an end load resistance Z. The pipe itself has some flow resistance R and the elasticity of the pipe leads to deformation governed by a parameter C. The meaning of the three parameters R, Z and C and the model variables P and Q is explained in detail in the text. We see that these five quantities are sufficient to explain the basic features of cardiovascular dynamics. Moreover, this system can be represented both as a differential equation model in Eqn. (1.4) and by an equivalent electronic circuit, Fig. 1.6.

This equation is rather complicated, and probably warrants a little closer examination before we move on. On the left-hand side we have a term for how the flow is changing in time $\frac{\partial u}{\partial t}$ and how the flow is being diffused in space $u \cdot \nabla u$. Remember that the operator ∇ is the gradient operator and that u, since it is measuring a velocity, is a vector. Hence, the term $u \cdot \nabla u$ measures how fast the fluid velocity is changing at a given location as a function of space, but not time. In Cartesian co-ordinates,

$$
u \cdot \nabla u = [u_x u_y u_z] \cdot \begin{bmatrix} \frac{\partial}{\partial x} \\ \frac{\partial}{\partial y} \\ \frac{\partial}{\partial z} \end{bmatrix} [u_x u_y u_z]
$$
$$
= \begin{bmatrix} u_x \frac{\partial u_x}{\partial x} + u_y \frac{\partial u_x}{\partial y} + u_z \frac{\partial u_x}{\partial z} \\ u_x \frac{\partial u_y}{\partial x} + u_y \frac{\partial u_y}{\partial y} + u_z \frac{\partial u_y}{\partial z} \\ u_x \frac{\partial u_z}{\partial x} + u_y \frac{\partial u_z}{\partial y} + u_z \frac{\partial u_z}{\partial z} \end{bmatrix},
$$

where $u = [u_x y_y u_z]$ is the Cartesian representation of the vector u. On the right-hand side we have one term which is due to the forcing exerted on the fluid by the change in pressure ∇P, and a second term which is the fluid diffusion $\nabla^2 u$ and depends on the viscosity ν. The second term measures the local smoothing out of the fluid flow. One expects that the velocity of a fluid must change smoothly in space, and that it should change more or less smoothly, depending on the fluid viscosity. Again, in Cartesian co-ordinates, the diffusion term can be written out in full as

$$\nabla^2 u = \begin{bmatrix} \frac{\partial^2 u_x}{\partial x^2} + \frac{\partial^2 u_x}{\partial y^2} + \frac{\partial^2 u_x}{\partial z^2} \\ \frac{\partial^2 u_y}{\partial x^2} + \frac{\partial^2 u_y}{\partial y^2} + \frac{\partial^2 u_y}{\partial z^2} \\ \frac{\partial^2 u_z}{\partial x^2} + \frac{\partial^2 u_z}{\partial y^2} + \frac{\partial^2 u_z}{\partial z^2} \end{bmatrix},$$

and each of these second spatial partial derivative terms $\frac{\partial^2}{\partial x^2}$ can be thought of as the curvature of the local velocity field. Hence, this term concerns the smoothness and the smoothing out of bumps in the velocity field as a function of viscosity.

The Navier–Stokes equation (1.1) can be thought of as a balance between the time rate of change $\frac{\partial u}{\partial t}$ and the spatial change $u \cdot \nabla u$, and the forcing due to applied pressure P and the fluid diffusion $\nabla^2 u$ (which is dependent on the fluid viscosity). Now, Eqn. (1.1) is rather general, but it already makes two important but reasonable assumptions about the nature of the fluid flow. This equation assumes that the fluid (blood) is incompressible and Newtonian. Essentially, incompressibility means that the flow is not in a regime that is likely to exhibit shock waves — something which is rather unlikely given the relatively slow velocity of blood flow. Secondly, a Newtonian fluid is one for which the relationship between fluid stress and strain is linear and can be directly quantified with the viscosity. Making this assumption means only that the fluid behaves like a "typical" fluid, irrespective of external agitation. Again, this is a reasonable assumption for blood.

Nonetheless, Eqn. (1.1) is still a little complicated and some additional simplification is useful (at the time of writing, there is an unclaimed \$US1,000,000 bounty for an adequate description of this equation and its generalisations). Therefore we will turn to something simpler by making further assumptions which will be applicable to our model in Fig. 1.5. Since the typical model blood vessel is cylindrical, we can exploit both the cylindrical symmetry (the blood vessel looks the same as it is rotated) and irrotationality (since there is no axial asymmetry). Equation (1.1) can then be reduced to a linear differential equation in time t and one spatial (longitudinal) direction x,

$$-\frac{dP}{dx} = RQ + L\frac{dQ}{dt}. \tag{1.2}$$

As before, P is the pressure term driving fluid flow. The variable Q is the rate of flow within a fixed volume, and R is a hydraulic resistance — the resistance to motion due to the fluid itself. The term L is derived from fluid density and is a measure of the fluid inertia.

Having reduced the Navier–Stokes equation to something a little more manageable, we are now half-way there. The other half of the solution comes from considering the blood vessel itself. Equation (1.2) concerns only the fluid flow (in this case blood) and does not consider the effect of the boundary — the blood vessel wall. In particular, the blood vessel wall is both elastic and somewhat porous. The elasticity, quantified by vessel wall compliance C, means

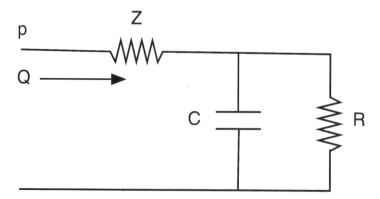

Figure 1.6: A linear analogue circuit model. The circuit shown consists of two resistors Z and R with values identical to the fluid resistance discussed in the text, and a capacitor C with capacitance equal to the arterial compliance of the blood vessel being modeled. The blood flow Q and the pressure P take the roles analogous to current and voltage (respectively) in the circuit.

that the blood vessel can deform to allow for greater or lesser blood flow, depending on the blood pressure. The porous boundary leads to a leakage term proportional to G and so the change in blood flow at the boundary is given by

$$-\frac{dQ}{dx} = GP + c\frac{dP}{dt}. \tag{1.3}$$

Combining Eqns. (1.2) and (1.3) finally leads to the arterial load model relating pressure P and blood flow Q

$$C\frac{dP}{dt} + \frac{1}{R}P = Q\left(1 + \frac{Z}{R}\right) + ZC\frac{dQ}{dt}, \tag{1.4}$$

where (following from before) C is the arterial compliance, R is the arterial flow resistance and Z is the end load (the aortic valve) resistance.

The interesting thing about Eqn. (1.4) is that it provides a model that relates flow to pressure and it is a first-order linear differential equation (admittedly, one of two variables) which can be represented exactly as an analogue circuit — see Fig. 1.6.

The key to model building in this example is to extract the salient feature, to discard the excess, and to be left with a model that is simple enough to use and complex enough to be useful. The model in Fig. 1.6 can be used to obtain a qualitative understanding of the function of certain parts of the circulatory system under various loads. For a given load P, it is a simple matter to solve Eqn. (1.4), and to obtain an expression for the flow. Of course, the

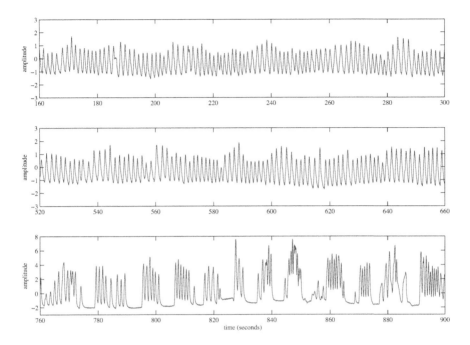

Figure 1.7: **An illustration of the onset of periodic breathing.** The figure depicts a continuous recording of respiratory effort (described in [35, 36]) of a sleeping human infant. The horizontal axis is time in seconds and the vertical axis is proportional to abdominal cross-sectional area. One can clearly see the onset of bursts of respiratory effort where the amplitude of respiration intermittently decreases to zero. Note that, during the regular respiration (up until $t = 660$) prior to onset of the periodic motion ($t = 760$ onwards), the amplitude of respiration is also periodically modulated.

model itself is rather simple, and to obtain such a model, many features of the underlying system have been omitted. Nonetheless, in this case, these are precisely the features which are not relevant to the specific interest we have in the system. For example, from the model we obtain, it is still possible to see how elasticity and hardening of the arterial walls can affect blood flow.

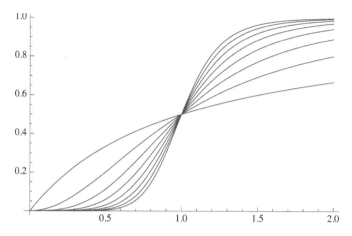

Figure 1.8: **The feedback function** $\frac{x^m}{1+x^m}$. The form of the feedback function (for $m = 1, 2, 3, \ldots, 8$) is plotted. In each case, the function is monotonic and bounded between 0 and 1 ($x > 0$).

1.3 Cheyne–Stokes Respiration

We now consider a second dynamical phenomenon in the human body. Figure 1.7 depicts a recording of respiratory effort (inferred from abdominal cross-sectional area measured via inductance plethysomnography — which will be discussed in more detail in Chapter 3) from a sleeping infant. The behaviour exhibited in this figure is perfectly normal and natural. Nonetheless, one can clearly observe two separate dynamical behaviours. For most of the recording we can see the regular rhythmic breathing one would expect — inhalation and exhalation. Admittedly, we do observe some fluctuation in the amplitude, but this is rather small. Then, something more dramatic happens. The respiratory effort periodically ceases. We see periodic fluctuations in the amplitude of respiration with the minimum amplitude dropping to zero. Effectively, the infant repeatedly ceases to breathe for periods of 2 to 3 seconds.

This behaviour can be understood based on a simple model of respiratory control feedback. Of course, all animals require oxygen and produce carbon dioxide as waste. That is, the level of carbon dioxide in the body increases in response to the bodily consumption of oxygen and, at the same time, respiration acts to remove the carbon dioxide from the body and from the blood-stream.

Let $c(t)$ denote the level of carbon dioxide in arterial blood, then

$$\frac{dc}{dt} = p - bVc, \tag{1.5}$$

where p is the constant rate of production of carbon dioxide by the body and bVc denotes the removal of carbon dioxide from the blood due to respiration. The term bVc is proportional both to the magnitude of the respiratory effort V and also the current carbon dioxide concentration c (that is, it is easier to remove carbon dioxide from the blood when there is more carbon dioixide in the blood — and vice versa). Suppose that there is some maximum achievable respiratory effort $v_{\max}$ and that V is itself a function of the level of carbon dioxide in the blood. But, the sensory receptors in the brain stem that control respiration do not function instantaneously, and so there is some delay in the feedback. Hence

$$V = V_{\max} \frac{c^m(t - T)}{a^m + c^m(t - T)}. \tag{1.6}$$

The form of the feedback Eqn. (1.6) is determined experimentally and it depends on the time delay T and also on two experimentally determined parameters a and m. Combining Eqns. (1.5) and (1.6) we obtain

$$\frac{dc}{dt} = p - bc(t)V_{\max} \frac{c^m(t - T)}{a^m + c^m(t - T)}.$$

But, this equation is a little misleading — symbolically it appears somewhat more complicated than what it actually is. To better elucidate the underlying meaning, we perform a process of nondimensionalisation: that is, we perform a specific substitution of variables aimed at making the actual mathematics clearer. In this case (following Murray [25]), let

$$x = \frac{c}{a}$$
$$\tilde{t} = \frac{pt}{a}$$
$$\tau = \frac{pT}{a}$$
$$\tilde{V} = \frac{V}{V_{\max}}$$

and then we obtain

$$\frac{dx}{d\tilde{t}} = 1 - \alpha x(\tilde{t}) \frac{x^m(\tilde{t} - T)}{1 + x^m(\tilde{t} - T)} \tag{1.7}$$
$$= 1 - \alpha x V(x(t - T)),$$

where

$$V(x) = \frac{x^m}{1 + x^m}$$
$$\alpha = \frac{abV_{\max}}{p}.$$

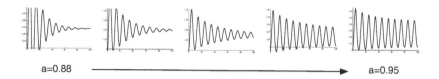

a=0.88   a=0.95

Figure 1.9: **Dynamical behaviour of Eqn. (1.8).** With $m = 10$ and $T = 4$, we vary a from 0.88 to 0.95 and observe a transition between stable fixed point and limit cycle — corresponding to that predicted by Eqn. (1.9).

By dropping the $\sim$, we can now re-write this as a somewhat simpler delay differential equation

$$\frac{dx}{dt} = 1 - \alpha x(t) V(x(t - T)), \tag{1.8}$$

where V is the velocity feedback function which in general will be monotonically increasing and bounded between 0 and $V_{\max}$ — this is illustrated in Fig. 1.8.

From Eqn. (1.8) we can perform some basic analysis to understand the dynamical behaviour of the system. First, we can observe that there exists a steady-state solution. That is, is there some fixed constant value that satisfies the Eqn. (1.8)? Suppose that there is, that is, $x(t) \rightarrow x_0$ where the value x_0 is fixed and does not change in time. Hence, if this solution exists, we expect that $\frac{dx}{dt}|_{x=x_0} = 0$ and $x(t) = x(t - T) = x_0$. Therefore

$$\frac{1}{\alpha x_0} = \frac{x_0^m}{1 + x_0^m}$$
$$1 + x_0^m - \alpha x_0^{m+1} = 0$$

Of course, the solution of this equation depends on the choice of m. Nonetheless, from Fig. 1.8 we see that the expression $\frac{x_0^m}{1+x_0^m}$ is monotonically increasing and sigmoidal. The expression $\frac{1}{\alpha x_0}$ is monotonically decreasing, and so there will exist a unique steady-state solution for $x_0 > 0$. From biological experiments there is reason to suppose that quite large values of m are actually appropriate. In Fig. 1.9 we take $m = 10$ and show that in this situation the system steady state x_0 can go from being stable to being unstable and that subsequently periodic solutions — corresponding to Cheynes-Stoke respiration — can arise. Analytically it is possible (with a bit more advanced mathematics) that if

$$\frac{dV}{dx}\bigg|_{x=x_0} > \frac{\pi}{2\alpha x_0 T} \tag{1.9}$$

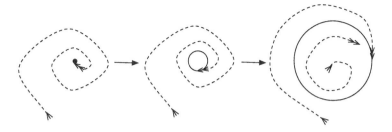

Figure 1.10: **The growth of a limit cycle.** From left to right we illustrate three distinct dynamical states. The system initially exhibits a stable fixed point and the system trajectory (dashed line) will converge to that. When the stable fixed point becomes unstable, a stable limit cycle forms around it (centre figure) and eventually grows (right). In the right-hand state, all trajectories (dashed lines) converge toward the limit cycle (stable circle) no matter whether they start inside or out. These states correspond exactly to what is observed in Fig. 1.9. In Fig. 1.9 we see the stable fixed point transition to a limit cycle as the dotted motion of a point goes from converging to a fixed-point to following a circular orbit.

then the steady state becomes unstable and periodic respiration occurs. Essentially, the stead- state x_0 transforms and becomes a limit cycle as depicted in Fig. 1.9 (and in Fig. 1.10). This type of change in dynamical behaviour will be a recurring theme in this text and is known as a *bifurcation* — the structure of the system's dynamical behaviour changes. Nonetheless, this is a difficult example to solve analytically. Rather than pursuing a derivation of Eqn. (1.9) we can happily resort to computational power. Simulations of the system dynamics for both $\frac{dV}{dx}\big|_{x=x_0} < \frac{\pi}{2\alpha x_0 T}$ and $\frac{dV}{dx}\big|_{x=x_0} > \frac{\pi}{2\alpha x_0 T}$ are illustrated in Fig. 1.9 and show the same change in dynamical behaviour.

Moreover, the same type of behaviour can be observed directly from computational models built from the data — this is illustrated in Fig. 1.11. In this figure we have generated computational (black box) models from the data in Fig. 1.7 and show that these models can also generate the same behaviour.

1.4 Summary

Biological systems change over time. In this chapter we introduced the concept of change — both as the system oscillates or otherwise moves in some consistent behaviour, and also as that behaviour may change from one time instant to another. To gain both a qualitative and quantitative understanding of how systems can and may change, we need to resort to models. In all cases,

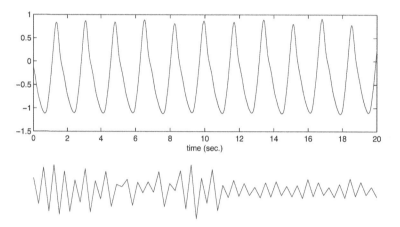

Figure 1.11: A computational model of respiration. The data from Fig. 1.7 (prior to the onset of Cheyne–Stokes respiration) was used to construct a predictive model of respiration. Essentially, the model is built from historical (past) data and attempts to predict the future values from the current state. These predictions are then iterated to produce the prediction shown in the upper panel. This is a simulation of respiration over 20 seconds, corresponding, in this case, to 11 breaths. One can observe that the model captures the fine, almost periodic (but not quite) behaviour. If we compute the amplitude of successive breaths (the lower line, computed over a longer duration), we obtain irregular variation in amplitude. These results indicate, consistent with our previous analysis, that this respiration sample is on the edge of the bifurcation from stable fixed point to periodic oscillation in breathing amplitude — and this is exactly what happens in the real data. A more detailed, model-based analysis of the transition is described in [33].

these models will be an abstraction of physical reality that somehow manage to capture the essential features of the underlying system. The models themselves can be mathematical, computational and even biological. The models can be either analytic or purely descriptive. Nonetheless, we aim to study the models to understand the dynamics of the systems from which the models are abstracted.

Glossary

Bifurcation The qualitative change in the system behaviour (from one stationary state to another) due to quantitative changes in specific system parameters. For example, altering a parameter may cause a periodic system to become chaotic.

Chaos A system that behaves in a manner that appears to exhibit randomness but is described entirely adequately without randomness. Mathematically, chaos is

1. Bounded,

2. Deterministic, and

3. Not periodic (or stationary).

Models A model is a simplification and abstraction of some observed thing. Typically, the models we consider will be mathematical descriptions of a specific real-world system. To be useful, such models must be simpler than the real world and yet contain sufficient complication to prove informative.

Qualitative Conversely, a qualitative description is one lacking numerical precision. Typically, a qualitative description relies on characterising the type of behaviour, but not the precise value.

Quantitative Something is quantitative if it can be measured and has specific numerical values assigned to it.

Stationary A stationary system is one that is not subject to any external influence or change. Hence, stationary systems will exhibit one consistent type of behaviour.

Exercises

1. Solve the circuit equation (1.4) for the case of steady-state pressure $P(t) = P_0$, and for when the aortic valve is closed (two separate cases). Describe the physiological meaning of each solution.

2. With pressure held constant, compute the time taken for blood flow to fall to within 1% of its asymptotic value. Use the (representative systemic arterial) values $Z_0 = 0.1$ mmHg-s/ml, $C_s = 1.5$ mL/mmHg and $R_s = 1.0$ mmHg-s/mL . With no heart dynamics, how long would it take to lose 1 L of blood? You may take the original rate of flow to be 5 L/min.

3. Consider the non-dimensionalised model of respiration given by

$$\frac{dx}{dt} = 1 - \alpha x(t) V(x(t - T)),$$

where $V(x) = \frac{x^m}{1+x^m}$ and $m = 1$ determined experimentally. What is the steady-state respiration? Show that perturbations to that steady state converge (i.e. that it is stable).

4. (a) Using the equation for arterial load $C_s \frac{dP}{dt} + \frac{1}{R_s} P(t) = Q(t) \left(1 + \frac{Z_0}{R_s} \right) + Z_0 C_s \frac{dQ}{dt}$, construct a model for regular sinusoidal forcing (i.e. an artificial pacemaker).

   (b) Hence, find the minimum flow under this model (HINT: The minimum applied force should be zero — but *Why?*).

   (c) Comment on your result.

5. Check that the dynamic behaviour observed in Fig. 1.9 is consistent with that predicted by Eqn. (1.8).

Chapter 2

Population Dynamics of a Single Species

2.1 Fibonacci, Malthus and Nicholson's Blowflies

The study of population dynamics — the changing population of a specie, or of species within an ecosystem — has a very long history. As human societies grew and people aggregated in denser and denser towns, cities and conurbations, the natural and vital question became "How many people will there be?". In the year 1202 AD , in his Latin text *Liber Abaci* [32], Leonardo of Pisa (aka Leonardo Pisano Bogolio or Leonardo Bonacci, or simply Fibonacci) asked the same question of rabbits: Suppose on the first day, I have one pair of adult rabbits. On each day, each pair of adults beget two infants. Infants grow to adulthood in one day, and all rabbits live forever. (Actually, the constraints proposed in 1202 AD were slightly different; This version is equivalent and conceptually easier to deal with.)

With modern algebra it is trivial to arrive at a solution. Let F_t be the number of *pairs* of rabbits (both adults and infants) on day t. Now the number of pairs of *adults* present on day t is given by F_{t-1} (remember that rabbits grow to adulthood in one day, and therefore those rabbits present on day $t-1$, will be adults by day t — if not sooner). So on day $t+1$ we will have all the pairs alive on day t, F_t (since all rabbits are immortal) plus a number of pairs of new babies equal to the number of adults on day t, F_{t-1}. Hence,

$$F_{t+1} = F_t + F_{t-1}.$$

Of course, in 1202 AD the method of deriving a solution prior to the invention of algebra was a little different: Basically, it was more wordy. The book *Liber Abaci* was actually an introduction to the West of many methods already well known to Egyptian, Middle Eastern and Indian mathematicians (including what would become known as the Fibonacci sequence, which arises from the solution of the rabbit problem described above). Nonetheless, it is possible to solve the problem arithmetically (with the same basic logic as that utilised in 1202 AD) — the process is illustrated in Fig. 2.1. Starting with one pair of infants on day 1, on day 2 we have one pair of adults, on day 3 we

Figure 2.1: **The growth of the Fibonacci sequence.** Black (filled) circles represent pairs of adults, white (open) circles represent the newly begot infants. Each line represents the population on a given day. Hence the number of circles on each line are the first ten Fibonacci numbers: $1, 1, 2, 3, 5, 8, 13, 21, 34$ and 55. This representation (where each pair of rabbits is represened by a circle) makes clear that each Fibonacci number is the sum of the preceding two terms.

have two pairs of rabbits (on adult one infant), on day 4 we have three pairs, on day 5 we have five, on day 6 we have eight, and so on. The sequence of numbers obtained from this process became known as the Fibonacci sequence, in honour of its Pisan populariser

$$0, 1, 1, 2, 3, 5, 8, 13, 21, 34, 55, 89, 144, 233, 377, 610, \ldots$$

Interestingly, among the other innovations presented to Europe in *Liber Abaci* were the numerals $0, 1, 2, 3, 4, 5, 6, 7, 8, 9$ themselves. Up until the twelfth century, Europe was still under the influence of the Roman numeral system: I, II, III, IV, V, ...

Nonetheless, the problem for us is how to describe the population growth indicated by the Fibonacci sequence. Plotting successive values of the sequence, it is clear that this formulation leads to a rapidly growing population — but how is this described mathematically? Actually, a (relatively) simple closed-form solution does exist. It is possible (although we won't be doing so here — if you're interested, consult the problems at the end of this chapter) to prove by induction that the Fibonnaci number F_t is given by

$$F_t = \frac{1}{\sqrt{5}} \left(\phi^t - (1 - \phi)^t \right),$$

where ϕ and $(1 - \phi)$ are the two solutions to the equation $x^2 - x - 1 = 0$ (note the similarity between this equation and the Fibonacci recurrence — that is the key to the proof by induction) and hence $(1 - \phi) = (-\frac{1}{\phi})$. For the sake of definiteness, if we take $\phi > 1/\phi > 0$, then $\phi = \frac{1+\sqrt{5}}{2} \approx 1.618$ is the so-called golden ratio. Hence, $1 - \phi \approx 0.618$ and so ϕ^t is exponentially increasing while $(1 - \phi)^t$ is oscillating and exponentially decreasing. For larger and larger t, the Fibonacci sequence looks more and more like a simple exponential — see Fig. 2.2.

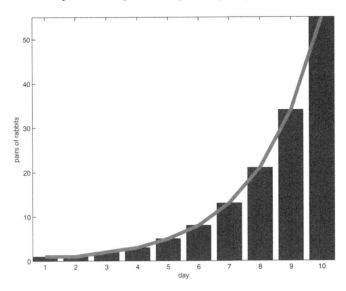

Figure 2.2: **Exponential growth of the Fibonacci sequence.** The first ten values of the Fibonacci sequence are plotted as a bar graph, and the explicit values computed according to $\frac{1}{\sqrt{5}}(\phi^t - (1 - \phi)^t)$ are shown superimposed.

In Europe towards the end of the eighteenth century, cities were growing rapidly and technological improvement within society saw vast improvement in the standard of living. At the beginning of the nineteenth century, the British theologian and scholar Thomas Malthus took these ideas and formulated a simple model of population growth. He observed (just as Fibonacci had done six centuries earlier) that populations tend to grow exponentially. Nonetheless, the available resources appear finite. That is, there is limited land and food to support burgeoning human societies. However, to claim that resources were finite was stricter than required. Malthus actually only claimed that available resources, or rather the technology to exploit them, would grow geometrically. That is, while population is increasing proportional to e^t, resources are increasing like t^n (for some fixed n). No matter what initial conditions one chooses (nor what exponent n), the exponential growth in population will outstrip the available resources.

Malthus' bleak conclusion was that population growth would continue unabated (unless through global moral restraint) to the point of crisis — the limiting factor for population growth would be war (over the control of and access to limited resources), pestilence (due to lack of sanitary water and other resources) or famine. Nonetheless, his ideas are mathematically very elegant. Let $N(t)$ be the population at a given time. Then, if births occur at some rate b and deaths at some rate d, the population growth rate increases at rate

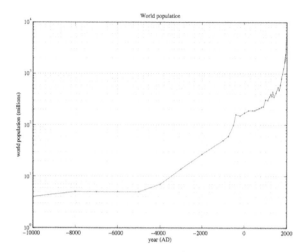

Figure 2.3: **Global population (logarithmic scale).** Total global population (estimated by various means) from Neolithic to modern times (10000 BC to 2000 AD).

bN and decreases at rate dN:

$$\frac{dN}{dt} = bN - dN \tag{2.1}$$

and hence $N(t) = N_0 e^{(b-d)t}$. The model is simple and the conclusions are startling, but nonetheless the results are well supported by data. In Fig. 2.3 the global population for the last 12 millennia has been increasing exponentially – with very minor deviation from the prediction of Eqn. (2.1). It is striking that one can actually pick out of Fig. 2.3 the transition from Stone Age to Bronze age around 4000–5000 BC and the industrialisation and urbanisation of the Industrial Revolution in the mid-nineteenth century.

Of course, Eqn. (2.1) actually says nothing about the factors limiting population. In 1838, Pierre François Verhulst extended the ideas of Malthus. Verhulst supposed that there was a fixed maximum population K — the *carrying capacity* — for a given system. When the population $N \ll K$, then the Malthusian relation (2.1) would hold approximately. However, as $N \to K$, the increase in population would slow, stop and then quickly reverse:

$$\frac{dN}{dt} = rN\left(1 - \frac{N}{K}\right). \tag{2.2}$$

Note that for $N \ll K$, the term $\left(1 - \frac{N}{K}\right) \approx 1$ and so the system behaves approximately Malthusively , $\frac{dN}{dt} \approx rN$. However, as $N \approx K$, the right-hand side approaches 0 and so population will stagnate. For $N > K$, the right-hand side becomes negative and the population will start to decrease. Hence, as

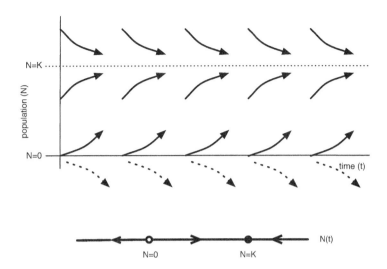

Figure 2.4: Dynamics of the logistic equation. The upper panel represents the dynamics of the Verhulstian population model $\frac{dN}{dt} = rN\left(1 - \frac{N}{K}\right)$ as a function of time. The two fixed points $N = 0$ and $N = K$ are shown with $N = K$ being attractive and $N = 0$ being repulsive. Below this we represent the same information as a one-dimensional state space.

in Fig. 2.4 we see that the system behaviour is constrained to increase if the value is small and decrease if it is large. Unfortunately, for Verhulst, Eqn. (2.2) has now become known not as the Verhulst equation, but as the logistic equation. Fortunately, for us, an analytic solution is obtainable.

From Eqn. (2.2) we can first find fixed points $\frac{dN}{dt} = 0$ corresponding to the stationary states of the system. $rN(1 - N/K) = 0$ implies that either $N = 0$ or $N = K$. That is, there are two stationary solutions — $N = 0$ (extinction) or $N = K$ (proliferation to the carrying capacity). It is also easy to see that the system is well bounded — that is, populations will not run away to infinity. If the population N is constrained by 0 and K ($0 < N < K$), then it is trapped by these two stationary states. Moreover, for $0 < N < K$, $\frac{dN}{dt} > 0$ and so the population will grow toward K (assuming $r > 0$, otherwise the converse is true). For $N > K$, we have that $\frac{dN}{dt} < 0$ and so the population will decrease to K. If $N < 0$, then $\frac{dN}{dt} < 0$ — fortunately, populations are almost always greater than zero. One can represent the expected dynamics as a one-dimensional phase space, as shown in the lower part of Fig. 2.4. Note that the two stationary states, which we will call *fixed points*, exhibit different dynamical behaviour — at least in their immediate neighbourhood. The fixed point at $N = 0$ is unstable — any solution close to $N = 0$ diverges from it. Meanwhile, the solution $N = K$ is stable — and any solution close converges to it.

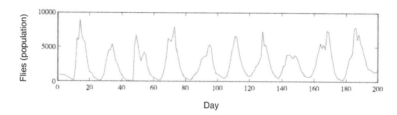

Figure 2.5: **Daily observed population of blowflies as recorded by Nicholson.**

The differential Eqn. (2.2) is nonlinear (since there is a term in N^2 on the right-hand side), but nonetheless it is still relatively easily solvable. For the time being though, we will obviate the need to do so by observing that the function

$$N(t) = \frac{N_0 K e^{rt}}{K + N_0(e^{rt} - 1)} \tag{2.3}$$

satisfies Eqn. (2.2) and so must be the unique solution to that differential equation. Note that $N_0 = N(0)$ and as predicted, $N(t) \to K$ and $t \to \infty$.

While the differential Eqn. (2.2) is nonlinear, it has an exact solution and the dynamical behaviour of the system is simple — convergence to the stable fixed point at $N = K$. In fact, we were able to deduce this by simply studying the sign of $\frac{dN}{dt}$, without actually solving the system. Moreover, the system state is completely defined by one scalar number, $N(t)$. The future values of the system are determined entirely by just the current value of N. Hence, one can imagine the solution as moving along a one-dimensional number line. This is the representation in the lower part of Fig. 2.4. Whatever point one chooses, its motion is constrained by that line. Since the system dynamics are not changing, the behaviour (at a particular value of N) will always be the same, irrespective of t. In the real world, this simple model has some explanatory power. Certainly, the global population data of Fig. 2.3 is consistent. But this is not the case in general. In Fig. 2.5 we illustrate a slightly more complex dynamical behaviour and will briefly mention the more sophisticated mathematics required to describe it.

In Fig. 2.5 we see the daily count of sheep blowflies recorded over a period of 200 days [12]. The data exhibits oscillators that are somewhat periodic — but not quite. The population goes through repeated booms and crashes without being exactly periodic. It is not, as predicted by Eqn. (2.2) a stable fixed point. In fact, for a one-dimensional system such as Eqn. (2.2) it is not possible to obtain bounded and nonperiodic behaviour such as this. As we showed in Fig. 2.4, if the system state is moving in one dimension, then it is constrained to go in the same direction for the same value of N every time. That is, if the system arrives at a value N at time t_1 and again later at $t_2 > t_1$, then at both times the system will progress and do exactly the same thing. Hence, it must also return to N at time $t_2 + (t_2 - t_1)$, $t_2 + 2(t_2 - t_1)$ and so on. That is, it must be periodic.

Therefore, to be able to explain the data of Fig. 2.5 we need a more complicated model. As Nicholson showed (and later discussed by Gurney and colleagues [12]), the blowfly population depends also on the larval population. Just as with Fibonacci's rabbits, the separate (intergenerational) populations of adults and children affect each other — so too here. Both adults and larvae are making simultaneous demands on food, and yet the current larval population depends on the adult population at some previous time. Hence, Eqn. (2.2) is modified as

$$\frac{dN}{dt} = rN(t)\left(1 - \frac{N(t - T)}{K}\right),\tag{2.4}$$

Because now the future behaviour depends not only on $N(t)$ but also on $N(t - T)$, the system is no longer constrained to exist in one dimension. That is, what happens in future is now no longer constrained to depend only on $N(t)$. Moreover, for reasons that are a little esoteric at this stage, this delay differential equation is now infinite dimensional. The future behaviour is not only two dimensional (for that too would not be enough to explain this observed behaviour) but it is infinite dimensional. The function $N(t)$ for all t in the interval $-T < t < 0$ is needed to describe the future. Similarly at any future time τ we need to know the value of N over an entire interval $\tau - T < t < \tau$ to be able to predict the future from Eqn. (2.4).

But it turns out that complicated (and therefore realistic) behaviour can also arise in much simpler models. In the next section we leave the world of the infinite dimensional, delay differential equations and return to one-dimensional difference equations: one-dimensional maps.

2.2 Fixed Points and Stability of a One-Dimensional First-Order Difference Equation

One-dimensional maps are the simplest dynamical models one can construct. In a one-dimensional map, time is discrete (so no more differential equations — for now) and the current system value x_t is a single number that depends only on the immediately preceding value: $x_t = f(x_{t-1})$. Because we are considering maps rather than differential equations, we are not so constrained by the restriction to one dimension as we were in the preceding section. With a differential equation the combination of one dimension and the fact that the dynamics are continuous means that the range of behaviour that can be observed is rather limited. With a difference equation, while still constrained in one dimension, the system state is only observed at discrete times and therefore can jump around in rather more complicated ways [21]. That is, since the state of the system at times other than these discrete observation times in unobserved, it is undefined.

Let us consider a simplification of the blowfly situation above. Suppose we have some species which reproduced in discrete generations. A good example is any animal with seasonal reproduction — if all the lambs are born in spring we can consider this as the unit time step and we are simply interested in predicting the number of sheep year-on-year. Essentially the logistic map is the discrete analogue of the Verhulstian logistic equation (Eqn. (2.2), except that this discrete map is capable of doing many more interesting things. The logistic map is given by

$$x_{t+1} = rx_t(1 - x_t), \tag{2.5}$$

where $0 \leq r \leq 4$ and the population x_t is expressed as a fraction, and so $0 \leq x_t \leq 1$. First, we shall observe that the restriction $r \geq 0$ is required to ensure that the population remains positive. We note in passing that $r > 4$ will also lead to negative population — because it will first cause $x_t > 1$.

As we did for the differential equations, the first step is to seek out the fixed points of the system. In the case of a map, the fixed points are solutions such that $x_{t+1} = x_t$. For the logistic map this means that the fixed points x_0 will satisfy $x_0 = rx_0(1 - x_0)$. Clearly, this equation has two solutions, $x_0 = 0$ and $x_0 = \frac{r-1}{r}$. Our next step is to look at what happens near the fixed points. Essentially we want to know whether if $x_t \approx x_0$, will x_{t+1} be closer to x_0 or further from it?

To achieve this in a way which is applicable to any one-dimensional map will require a little work — while there may be quicker ways to answer this question for the logistic map, let us take the slower and more general path so that it can then be applied equally to other systems. But first we will consider one specific example to motivate the method. Consider the fixed point $x_0 = 0$ of the logistic map. For $x_t \approx x_0$, the logistic map looks mostly like $x_{t+1} = rx_1$ (since if $x_t \approx 0$, then x_t^2 will be even smaller). Hence, if $x_t \approx x_0$, then $x_{t+1} \approx rx_t$. If $r < 1$ (actually if $|r| < 1$), then x_{t+1} will be closer to x_0 than x_t was. If $r > 1$, then the reverse is true. Hence, if $|r| < 1$, then the fixed point at $x_0 = 0$ is stable, otherwise it is unstable.

That is the basic principle. Now, we can extend it to any one-dimensional map $x_{t+1} = f(x_t)$ with a fixed point x_0 such that $x_0 = f(x_0)$. First we perform a Taylor series expansion to linearise about the fixed point. That is, we make the following series expansion

$$f(x_t) = f(x_0) + f'(x_0)(x_t - x_0) + \frac{1}{2!}f''(x_0)(x_t - x_0)^2 \tag{2.6}$$

$$+ \frac{1}{3!}f'''(x_0)(x_t - x_0)^3 + \dots$$

If $x_t \approx x_0$, then $(x_t - x_0)$ will be small and (provided the higher-order derivatives are bounded), higher-order terms will vanish. The dominant term be-

TABLE 2.1: **Stability of a fixed point of a 1-D map.**

	$\|f'(x_0)\| < 1$	$\|f'(x_0)\| > 1$
$f'(x_0) > 0$	stable node	unstable node
$f'(x_0) < 0$	stable focus	unstable focus

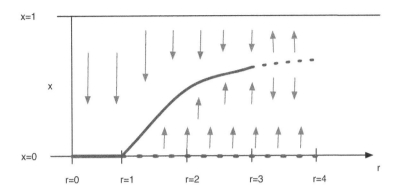

Figure 2.6: **Bifurcation diagram for the logistic map.** The horizontal axis is r and so the fixed dynamics of the system in a given state may be considered (as fixed r) a one-dimensional vertical slice through this diagram. For $r < 1$, there is only one fixed point in the interval $0 \le x \le 1$ and it is stable. For $1 < r < 3$, the stable fixed point moves away from $x = 0$ (actually, this is the second fixed point) and the fixed point at $x = 0$ becomes unstable. For $r > 3$, both fixed points are unstable — and yet we have already observed that for $r < 4$ if $0 \le x_t \le 1$, then the iterates of x_t will always be between 0 and 1. Something more interesting is happening for $r > 3$.

comes the first derivative and we can neglect the rest

$$x_{t+1} \approx x_0 + f'(x_0)(x_t - x_0) \tag{2.7}$$

(and remember that $f(x_0) = x_0$). The question is now: is x_{t+1} closer to x_0 than x_t, or not? It is easy to see that $x_{t+1} - x_0 = f'(x_0)(x_t - x_0)$. Hence, if $|f'(x_0)| < 1$, then x_{t+1} will be closer to x_0. If $|f'(x_0)| > 1$, then x_{t+1} will be more distant. That is, if $|f'(x_0)| < 1$, the fixed point is stable (since $x_t \to x_0$); if $|f'(x_0)| > 1$, then it is unstable (as $|x_t - x_0| \to \infty$). Moreover, if $f'(x_0) < 0$, we can see that the successive points will be alternatively larger and smaller than x_0 (that is, the sign of $(x_t - x_0)$ will alternate). Conversely, if $f'(x_0) > 0$, then the sign of $(x_t - x_0)$ will not change. Note that for $|f'(x_0)| = 1$, we cannot say anything about the stability of the fixed point (we need to look at the higher-order derivatives) — and similarly for $f'(x_0) = 0$ when considering whether the solution oscillates.

Technically, a fixed point that oscillates is a *focus* (plural *foci*); one that does not is a *node*. Table 2.1 summarises the possible classifications one can make based on $f'(x_0)$. Return to the logistic map (2.5). Setting $f(x) = rx(1-x)$, we obtain $f'(x) = r(1-2x)$ and for the two fixed points $x_0 = 0$ and $x_0 = \frac{r-1}{r}$, we have $f'(0) = r$ and $f'\left(\frac{r-1}{r}\right) = 2 - r$. Hence, the fixed point at $x_0 = 0$ is stable if $0 < r < 1$ (we only consider $r > 0$) and unstable for $r > 1$. The fixed point at $x_0 = \frac{r-1}{r}$ is stable for $1 < r < 3$ and unstable for $r > 3$. We can now produce a bifurcation diagram to illustrate the location and stability of the fixed points (on the interval $0 \le x \le 1$ as a function of r) — this is illustrated in Fig. 2.6.

2.3   The Cobweb Diagram

We can now attempt to understand the complicated dynamics of the logistic map that may arise, particularly for $3 < r < 4$ — graphically. In Fig. 2.7 we illustrate a single trajectory of the logistic map, for $r = 2.5$, and we see that the trajectory converges to a fixed point. This result is exactly what was predicted in the previous section: since for $r = 2.5$ we have $1 < r < 3$, and hence the fixed point $\frac{r-1}{r}$ is stable.

Nonetheless, to arrive at this conclusion by graphical means we can take a few more steps. Let $f(x) = rx(1-x)$ denote the logistic map, as defined previously. The diagram in Fig. 2.7 is a representation of successive pairs of points (x_t, x_{t+1}). Any point (x_t, x_{t+1}) will be somewhere on the parabola $y = rx(1-x)$. Here, we now use x and y to denote the horizontal and vertical coordinates, as usual. To find the next point (x_{t+1}, x_{t+2}), one can adopt the following geometric reasoning:

1. Project horizontally from the point (x_t, x_{t+1}) to the line $y = x$. Since the horizontal translation has left the vertical coordinate unchanged, this point is (x_{t+1}, x_{t+1}).

2. Project vertically from the point (x_{t+1}, x_{t+1}) to the parabola $y = rx(1 - x)$. Since the vertical translation has left the horizontal coordinate unchanged, this point is $(x_{t+1}, rx_{t+1}(1 - x_{t+1}))$, which, of course, is (x_{t+1}, x_{t+2}).

3. Increment t and repeat.

When this procedure is repeated for various values of r, the rich range of dynamics of the logistic map emerge. In Fig. 2.8 the procedure illustrated in Fig. 2.7 has been repeated several times. As r increases, the dynamics change from the single fixed point depicted in Fig. 2.7 to a period 2 orbit — that is, the system alternates between two points. A period 2 (and also, with a bit more complication, higher-order period) orbit can also be verified analytically. A period 2 orbit is a pair of points $a_x \ne x_2$ such that $x_2 = f(x_1)$ and, vice

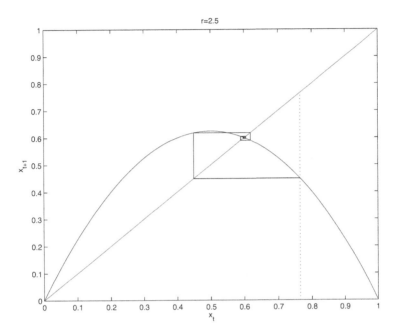

Figure 2.7: **Cobweb diagram of the logistic map.** The cobweb diagram here illustrates the dynamics of the logistic map for $r = 2.5$ when the system exhibits a single stable fixed point $x_0 = \frac{1.5}{2.5} = 0.6$. The downward-facing parabola indicates the logistic map, the diagonal line is the identity. The initial condition (chosen simply for illustration) is marked as a dotted vertical line. From the intersection of this initial condition and the parabola, we obtain the first iterate of the logistic map. This is then projected (and reflected) through the identity line to obtain the next iterate at the next intersection with the parabola. Each iteration is shown as two line segments with end points intersecting at a right angle (an "L" shape, possibly reflected and/or rotated). The algorithm is described in the text.

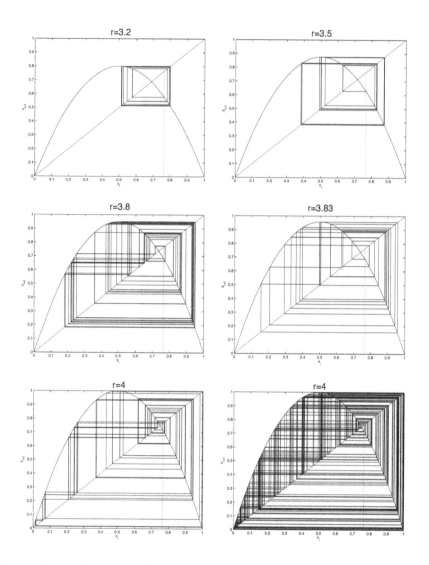

Figure 2.8: **Further cobweb diagrams of the logistic map.** We have
selected five further values of the logistic map and repeated the procedure
described in Fig. 2.7. For $r = 3.2$, the system is period 2 — switching
between two values (one around 0.5 and one around 0.78). For $r = 3.5$, the
system is period 4. For $r = 3.83$, the system is period 3. In this case it is
a little harder to see, because it takes a rather long path before reaching the
periodic orbit. Nonetheless, by following the orbits starting at a value around
0.15, it is clear that the system switches continually between three distinct
values. Finally, for both $r = 3.8$ and $r = 4$, the system does not settle to any
fixed values but continues to change — this is chaos. The final two panels in
this figure are both for $r = 4$; one is just extended for a longer time period to
indicate that the system really does not settle down.

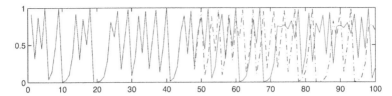

Figure 2.9: Sensitivity to initial conditions We depict exponential divergence of two initially close states (differing only by 64-bit computational machine precision) of the logistic map $r = 4$. After about 50 iterations, the divergence of the two trajectories is easily visible. After only a very few further steps, the trajectories are complete independent.

versa,$x_1 = f(x_2)$. From period 2, the system then goes on to exhibit period 4, then period 8 (this sequence, not surprisingly, is called a *period doubling bifurcation*).

Eventually, we reach values of r (in Fig. 2.8, $r = 3.8$ and $r = 4$) such that the system no longer eventually settles down to a periodic orbit. Mathematically, we observe that the system f has trajectories x_t with the following three properties:

- **Bounded:** The system does not get infinitely large or small. We can find numbers (bounds) m and M such that $m < x_t < M$ for all x_t.

- **Deterministic:** That is, it can be described and predicted exactly by a precise mathematical equation. The system is not random, nor is its behaviour influenced by noise.

- **Aperiodic:** The system will never repeat. For (almost[1]) all initial conditions, if we iterate the initial condition, we will never find $x_t = x_t + \tau$ for any t and $\tau > 0$.

When taken together, a trajectory x_t that exhibits these three properties is said to be *chaotic*. A system f is chaotic if almost all of its trajectories are. We need to keep the proviso "almost all" because there will be some trajectories which are exactly periodic. But, these will all be unstable, hence starting arbitrarily close to an unstable periodic trajectory will still produce a chaotic trajectory. In a rigorous sense, by "almost all" we mean that the chance of finding a periodic solution by guessing is 0.

There are actually several alternative definitions for what constitutes chaos. For our purposes, a deterministic system exhibiting bounded aperiodic trajectories (almost always) is a perfectly good one. One other definition worth mentioning is sensitivity to initial conditions. That is, if we start out with two

---

[1]Here, we mean that the chance of choosing a point which is periodic has probability 0. Such points will exist, but they are extremely rare.

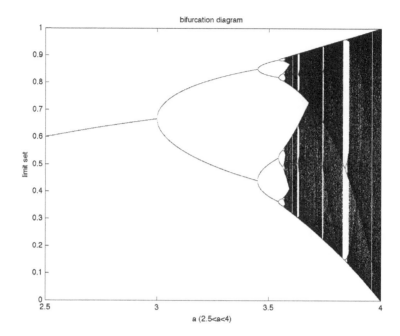

Figure 2.10: **Bifurcation diagram for the logistic map** $x_{t+1} = rx_t(1 - x_t)$. For $3 \leq r \leq 4$ we compute iterates of random initial conditions. The procedure is described in detail in the text. The result clearly shows that the stable fixed point bifurcates repeatedly to higher and higher order periodic orbits — with the period doubling on each occasion. Chaotic dynamics follows with various brief windows of periodic dynamics. For $0 < r < 3$ the dynamics of the system is described completely accurately by the bifurcation diagram in Fig. 2.6.

states very close to each other (as close as we care to take), then, over time the difference between those two states will increase — and it will increase exponentially (see Fig. 2.9).

In Fig. 2.6 we constructed a sketch of the bifurcation behaviour of the logistic map. Unfortunately, for $r > 3$ both fixed points became unstable and it was no longer clear what type of dynamics should be expected. To answer this question most directly, we can resort to computational methods. For a fixed value of r, we can choose initial conditions (actually one is enough) in the interval $0 < x_0 < 1$ and compute iterates of x_0: $x_{t+1} = f(x_t)$. In Fig. 2.10 we plot $(x_{10000}, x_{10001}, \ldots, x_{20000})$ for $2 < r < 4$. Of course, the reason for rejecting the first 10000 iterates is that we must ensure that we are approaching the behaviour of $t \to \infty$. Obviously, $t \ll \infty$ but, as the rapid (exponential) divergence of initial conditions in Fig. 2.9 shows, 10000 steps is far more than needed.

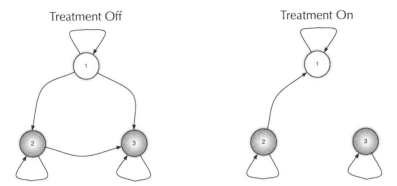

Figure 2.11: Model of Intermittent Androgen Suppression (IAS) treatment. Cells are categorised as either responsive to treatment (state 1); unresponsive, but recoverable (state 2); or, nonreversibly unresponsive (state 3). During the treatment-free state cells can progress from state 1 to 2 or 3 (and from state 2 to 3). During treatment, cells in state 2 can return to state 1. The rates of these various exchanges are determined by the parameters $d_{i,j}^{0,1}$ in Eqns. (2.8) and (2.9).

2.4 An Example: Prostate Cancer

A recent mathematical model of hormone treatment therapy for prostate cancer [14] was developed by Hirata and colleagues. Cancer of the prostate is (usually) slow growing and typically treated with courses of androgen-suppression drugs. Because the cancerous cells in the prostate respond to the same regulatory mechanism as regular cells, it is possible to control the growth of such cells. This growth is regulated by the presence of testosterone (which, among other things, is an androgenic steroid) and hence androgenic-suppression treatment will limit the growth of these testosterone-dependent cells. Conventional wisdom asserts that continuous application of these androgen suppressors, Continuous Androgenic Suppression (CAS), will most effectively limit the growth of the targeted cancerous cells. However, over time, the efficacy of these chemicals and therefore of the treatment may decrease and some patients will experience relapse. However, the growth of cancerous cells is, of course, a dynamical process and the question is whether continuous application of androgenic suppressors is the most effective course.

Following [14], we can consider three distinct populations of cells: (1) androgenic dependent, (2) reversibly androgenic independent and (3) irreversibly androgenic independent. We represent these populations as x_1, x_2 and x_3. The total population, of course, is $x_1 + x_2 + x_3$. One can then obtain two sets of difference equations, one corresponding to when treatment with androgenic

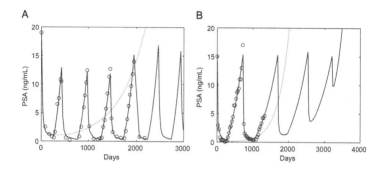

Figure 2.12:   Results of application of Intermittent Androgen Suppression (IAS) treatment. The presence of antigens (in this case, an indicator of cancer) for two real patients (circles) from model predictions with IAS (heavy line) and CAS (lighter line). (Figure taken with permission from [14].)

suppression is applied and one when it is not. During treatment:

$$
\begin{pmatrix} x_1(t+1) \\ x_2(t+1) \\ x_3(t+3) \end{pmatrix} = \begin{pmatrix} d_{1,1}^1 & 0 & 0 \\ d_{2,1}^1 & d_{2,2}^1 & 0 \\ d_{3,1}^1 & d_{3,2}^1 & d_{3,3}^1 \end{pmatrix} \begin{pmatrix} x_1(t) \\ x_2(t) \\ x_3(t) \end{pmatrix}
\tag{2.8}
$$

and, during nontreatment,

$$
\begin{pmatrix} x_1(t+1) \\ x_2(t+1) \\ x_3(t+3) \end{pmatrix} = \begin{pmatrix} d_{1,1}^0 & d_{1,2}^0 & 0 \\ d_{2,1}^0 & d_{2,2}^0 & 0 \\ d_{3,1}^0 & d_{3,2}^0 & d_{3,3}^0 \end{pmatrix} \begin{pmatrix} x_1(t) \\ x_2(t) \\ x_3(t) \end{pmatrix}.
\tag{2.9}
$$

The parameters $d_{i,j}^0$ during nontreament and $d_{i,j}^1$ during treatment are patient dependent and characterise both the aggressiveness of this specific cancer in that individual and also the patient's responsiveness to treatment — this is depicted in Fig. 2.11.

It is possible to estimate all the parameters of Eqns. (2.8) and (2.9) from actual clinical data and then build a model for an individual patient. This model can then be used to both predict the efficacy of IAS and CAS and also to design a most appropriate response. In Figure 2.12 we show (from [14]) real data from two patients (representative of a much larger sample). IAS therapy benefits the patient by preventing relapse (in panel (A) or delaying it (in panel (B)). The model we introduce is essentially two sets of linear difference equations which one can switch between. In the next section, we consider the more general case of a system of nonlinear equations.

2.5 Higher-Dimensional Maps

We have already seen how to characterise the stability of fixed points for one-dimensional maps. But what about higher dimensions? We will consider, as a concrete example, the two-dimensional Hénon map (because it is simple and very widely known — not because of its biological relevance):

$$x_{t+1} = y_t + 1 - ax_t^2 \tag{2.10}$$

$$y_{t+1} = bx_t \tag{2.11}$$

(a and b are parameters — we'll return to them later) as a special case of

$$x_{t+1} = f(x_t), \tag{2.12}$$

where $f : \mathbf{R}^d \to \mathbf{R}^d$ is a d-dimensional map.

First we will state the general results and then apply them to the specific example of the Hénon map. It should come as no great surprise that one can find fixed points by simply computing solutions x_0 of $x_0 = f(x_0)$. The question remains: what about the stability of the fixed points?

Just as before, we can linearise about the fixed point and express the local dynamics in terms of a Taylor series — only this time we need to use d-dimensional matrices. For $x_t \approx x_0$,

$$x_{t+1} = f(x_0) + (x_t - x_0)^T \nabla f(x_0) + \dots, \tag{2.13}$$

where $(\cdot)^T$ is the vector transpose operator and $\nabla f(x)$ is the gradient of f

$$\nabla f(x) = \begin{bmatrix} \frac{\partial f_1}{\partial x_1} & \frac{\partial f_2}{\partial x_1} & \cdots & \frac{\partial f_d}{\partial x_1} \\ \frac{\partial f_1}{\partial x_2} & \frac{\partial f_2}{\partial x_2} & \cdots & \frac{\partial f_d}{\partial x_2} \\ \vdots & \vdots & \ddots & \vdots \\ \frac{\partial f_1}{\partial x_d} & \frac{\partial f_2}{\partial x_d} & \cdots & \frac{\partial f_d}{\partial x_d} \end{bmatrix} \tag{2.14}$$

We'll stop the expansion in Eqn. (2.13) at the first derivative because the notation involved in the higher-order derivatives gets rather messy — and we don't really need it. We can now re-write Eqn. (2.13) as

$$x_{t+1} = x_0 + (x_t - x_0)A \tag{2.15}$$

and we can then perform a change of variables by diagonalising the matrix $A = \nabla f(x_0)$. A matrix A is said to be diagonalised if we can re-write A as $A = PDP^{-1}$, where P is invertible and D is a diagonal matrix (that is, all elements not on the diagonal are zero). A matrix A is diagonalisable if it has full rank — that is, if it has as many distinct eigenvalues as it has rows. Moreover, suppose that the matrix A has eigenvalues $\lambda_1, \lambda_2, \dots, \lambda_d$

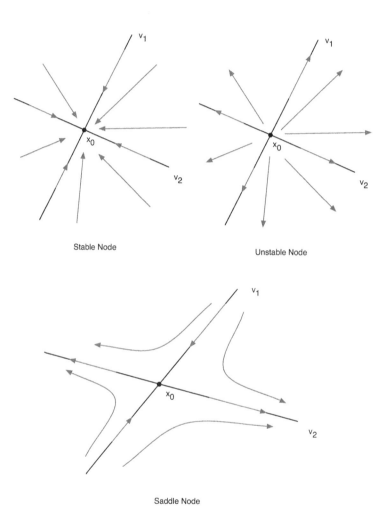

Figure 2.13:   Stability of a fixed point of a two-dimensional map. In two dimensions, the fixed point x_0 will have two eigenvalues λ_1 and λ_2 with associated eigenvectors v_1 and v_2. If $|\lambda_{1,2}| < 1$, the fixed point is a stable node. If $|\lambda_{1,2}| > 1$, the fixed point is an unstable node. If $|\lambda_1| < 1 < |\lambda_2|$, then the fixed point is a saddle, with a stable direction v_1 and an unstable direction v_2.

($\lambda_1 > \lambda_2 > \ldots > \lambda_d$ — complex conjugate pairs are OK too (we'll come to them in a moment) and corresponding eigenvectors $v_1, v_2, \ldots, v_d$.

Now, we can diagonalise A in the following way:

$$A = \begin{bmatrix} v_1 & v_2 & \cdots & v_d \end{bmatrix} \begin{bmatrix} \lambda_1 & 0 & \cdots & 0 \\ 0 & \lambda_2 & \cdots & 0 \\ \vdots & \vdots & \ddots & \vdots \\ 0 & 0 & \cdots & \lambda_d \end{bmatrix} \begin{bmatrix} v_1 & v_2 & \cdots & v_d \end{bmatrix}^{-1}, \tag{2.16}$$

where the eigenvectors are represented here as column vectors — and their concatenation produces a square matrix. Remember that eigenvalues can be found by solving the equation $\det(A - \lambda I) = 0$ and the eigenvectors v satisfy $Av = \lambda v$. Now, Eqn. (2.15) can be re-written as

$$x_{t+1}P = x_0 P + (x_t P - x_0 P)D \tag{2.17}$$

and so we can perform the change of variables $\xi_t = x_t P$ and we have

$$\xi_{t=1} = \xi_0 + (\xi_t - \xi_0)D \tag{2.18}$$

where, since the matrix D is diagonal, the original d-dimensional map (2.13) is now a system of d one-dimensional maps (2.18). Of course, the stability of each direction is governed by the corresponding eigenvalue (just as described in Table 2.1: if $|\lambda_i| < 1$, then the system (2.18) is stable on its i-th coordinate. Equivalently, remember that the change of variables $\xi = xP$ amounts to a rotation and dilation according to the matrix P and so we have

- If $|\lambda_i| < 1$, then the system (2.17) is stable at the fixed point x_0 in the direction v_i.

- If $|\lambda_i| > 1$, then the system (2.17) is unstable at the fixed point x_0 in the direction v_i.

Hence, a fixed point is only stable (and is called a *stable node*) if $|\lambda_i| < 1$ for $i = 1, 2, \ldots d$. A fixed point is unstable (and is called an *unstable node*) if $|\lambda_i| < 1$ for $i = 1, 2, \ldots d$. If the fixed point has some eigenvalues for which $\|\lambda_i\| > 1$ and some for which $|\lambda_i| < 1$, then it is a saddle point — this means that there are some directions for which the fixed point is attractive and some for which it is repulsive. Points nearby to a saddle will approach it and then move away from it. Only if the point lies exactly on the stable direction(s) (with no part in an unstable direction) will the point reach the fixed point — and then stay there. The three possible scenarios — stable node, unstable node and saddle — are illustrated for dimension 2 in Fig. 2.13.

Algebraically, there is one important possibility that we have neglected — the eigenvalues could occur as a complex conjugate pair. If this is the case, then it is no longer possible to directly separate the system into d independent one-dimensional systems. If we just focus on the two directions corresponding

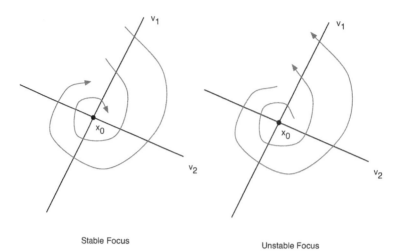

Stable Focus   Unstable Focus

Figure 2.14: Stability of a fixed point of a two-dimensional map — complex eigenvalues. In two dimensions, the fixed point x_0 will have two eigenvalues λ_1 and λ_2 with associated eigenvectors v_1 and v_2. If λ_1 and λ_2 form a complex conjugate pair, then the system will exhibit a focus. If $|\lambda_{1,2}| < 1$, the fixed point is a stable focus. If $|\lambda_{1,2}| > 1$, the fixed point is an unstable focus.

to the complex conjugate eigenvalue pair, then we may suppose that we have the system

$$x_{t+1} = x_0 + (x_t - x_0)A_{11} + (y_t - y_0) * A_{12} \tag{2.19}$$

$$y_{t+1} = y_0 + (x_t - x_0)A_{21} + (y_t - y_0) * A_{22}, \tag{2.20}$$

where the matrix

$$A = \begin{bmatrix} A_{11} & A_{12} \\ A_{21} & A_{22} \end{bmatrix}$$

has a complex conjugate pair of eigernvalues λ and $\overline{\lambda}$. In this case, it is more appropriate to transform (x, y) into polar coordinates $x_t = r_t \cos \theta_t$ and $y_t = r_t \sin \theta_t$, and then the stability of the system can be determined from the magnitude ρ of the eigenvalues $\lambda = \rho e^{i\theta}$ and $\overline{\lambda} = \rho e^{-i\theta}$. If $\|\lambda\| = |\rho| < 1$, the system is stable, if $\|\lambda\| = |\rho| > 1$, then it is unstable. However, unlike the case for a stable node depicted in Fig. 2.13, the stable trajectories rotate as they converge to the fixed point. In this case the system is either a *stable focus* or an *unstable focus*, as depicted in Fig. 2.14.

We will now finish with the example system with which we opened this section (2.11):

$$x_{t+1} = y_t + 1 - ax_t^2$$

$$y_{t+1} = bx_t,$$

where it is usual to set $b = 0.3$. A bifurcation resulting in the onset of chaos can be described as a function of b. The fixed points (x_0, y_0) will occur when

$$x_0 = y_0 + 1 - ax_0^2,$$
$$y_0 = bx_0.$$

That is (upon substituting the second equation into the first and rearranging the result), $ax_0^2 + (1 - b)x_0 - 1 = 0$ which implies we need

$$x_0 = \frac{-(1 - b) \pm \sqrt{(1 - b)^2 + 4a}}{2a},$$

and fixed points will only exist if $a > -\frac{1}{4}(1 - b)^2$ and they will occur as a pair. The stabilty of the fixed points can be determined by examining

$$\nabla f = \begin{bmatrix} -2ax & 1 \\ 0 & b \end{bmatrix},$$

where

$$f = \begin{bmatrix} y + 1 - ax^2 \\ bx \end{bmatrix}.$$

To solve this satisfactorily, one needs either algebraic prestidigitation or some numerical values. Nonetheless, substituting the value for x_0 into this matrix, we can find the eigenvalues $\lambda_{1,2}$ as the solution to

$$\det (A - \lambda I) = 0,$$
$$A = \begin{bmatrix} (1 - b) \pm \sqrt{(1 - b)^2 + 4a} & 1 \\ 0 & b \end{bmatrix};$$

hence

$$((1 - b) \pm \sqrt{(1 - b)^2 + 4a} - \lambda)(b - \lambda) - 0.1 = 0,$$

which implies

$$\lambda_1 = (1 - b) \pm \sqrt{(1 - b)^2 + 4a},$$
$$\lambda_2 = b.$$

The fixed point at $x_0 = \frac{-(1-b)+\sqrt{(1-b)^2+4a}}{2a}$ will be a stable node if $|(1-b)\pm \sqrt{(1-b)^2 + 4a}| < 1$ and the fixed point at $x_0 = \frac{-(1-b)-\sqrt{(1-b)^2+4a}}{2a}$ will be a stable node if $|(1-b) - \sqrt{(1-b)^2 + 4a}| < 1$. Otherwise, they are saddles. If we take $b = 0.3$, we can compute the required values. The bifurcation diagram (for $b = 0.3$) for the x-coordinate of the Hénon map is shown in Fig. 2.15.

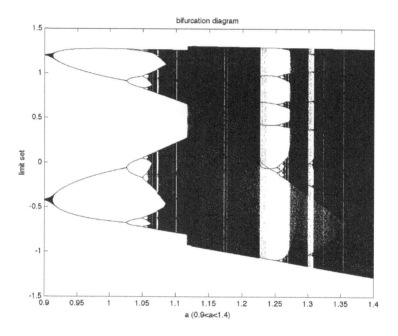

Figure 2.15: Bifurcation diagram for the Hénon map. For $0.9 \leq r \leq 1.4$ we compute iterates of random initial conditions. The procedure is described in detail in the text. The result clearly shows that the stable fixed point bifurcates repeatedly to higher and higher order periodic orbits — with the period doubling on each occasion. Chaotic dynamics follow with various brief windows of periodic dynamics.

2.6 Period-Doubling Bifurcation in Infant Respiration

We now conclude this chapter with an example of a real system exhibiting exactly the type of period doubling bifurcation depicted in Fig. 2.10. Moreover, the system is an inherently continuous one — the human respiratory system. Nonetheless, by performing a simple transformation, we extract a discrete system and show that the bifurcation patterns one observes are equivalent to the period doubling bifurcation in the logistic map.

For many systems that are changing continuously, the change one observes can be largely characterised by the variation from one orbit to the next. In the case of human respiration (and many, many other systems — but let's stick to one example), the system exhibits repeated cycles, and predicting what happens over one cycle is not particularly challenging. Inspiration (breathing in) will always (well, almost always — with at most one exception per person) be followed by exhalation (breathing out). What is more interesting is the variation *between* these repeating cycles. What happens on successive breaths? Is there a relationship between the amplitude of this and the next breath? In fact, we have already seen evidence of just such a dynamic in the guise of Cheyne-Stokes respiration (that is, the so-called "periodic" breathing[2]). Mathematically, one can construct the relationship between the apparently periodic continuous system and the discrete as shown in Fig. 2.16. The construction here is known, in honour of its originator Henrí Poincaré, as the Poincaré section.

2.7 Summary

The rabbit problem posed at the start of the thirteenth century by Fibonacci shows us how the description of the structure of a dynamically changing system can be used to describe the nature of that change. Either differential equations or the difference equations describing a system can be used to determine the fixed points of that system. In this chapter we focussed on the fixed points of difference equations (maps), because this is the simplest and easiest to deal with. From the stability of these fixed points we are able to infer the behaviour of the whole system. In the coming chapters we extend this idea to higher-dimensional system and to continuous systems: systems of *differential* equations. But first, in the next chapter we consider a more

---

[2]Of course, all breathing is periodic, but what makes periodic breathing periodic is that the amplitude changes periodically.

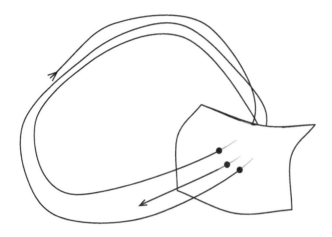

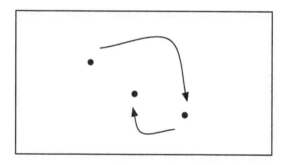

Figure 2.16: A Poincaré section. When studying the dynamics of a continuous system (the upper panel), it is often convenient to reduce it to a discrete one. In this example the continuous system is almost periodic, the trajectory orbits around, almost periodically. Hence, what is of most interest to us is the variation from that periodic repeating pattern. To examine this we construct a hyper-plane (if the original continuous dynamics are represented in n-dimensions, then the discrete map is obtained by the $(n-1)$-dimensional plane, a hyper-plane) and plot the intersection of the continuous dynamics and this plane. The hyper-plane is chosen so that it is orthogonal to the dynamics, and therefore the successive intersections of a trajectory and this plane are points in $(n-1)$-dimensional space. The dynamics of the system are then described by the successive location of these points.

basic engineering problem — the physical measurement of a signal from the underlying deterministic dynamical system.

Glossary

Bifurcation diagram Changing dyamics of the system plotted as a function of a change in one of the system parameters. That is, the bifurcation diagram illustrates the change in the behaviour of the system with a change in one of the parameters of the system.

Chaos A system is chaotic if almost all its trajectories are chaotic. By almost all, we mean that the chance of randomly choosing a non-chaotic trajectory in a chaotic system is 0.

Cobweb diagram A geometric approach to understanding the stability of (particularly) one-dimensional maps. By plotting the map as a function of state space, one can geometrically iterate the system to understand the stability of particular trajectories.

Fixed point A fixed point of a discrete map f is a point x_0 such that $f(x_0) = x_0$. Loosely speaking, it is a state of the system such that if the system is in that state, it will stay in that state. The interesting question to ask is what will happen *near* the fixed point.

Focus A fixed point is a focus if trajectories converge to it or diverge from it via rotation (orbiting) around the fixed point.

Node A fixed point is a node if trajectories converge to it or diverge from it directly.

Period-doubling bifurcation A particular type of bifurcation. The system stability increases from a stable fixed point to period-2, then period-4 and so on until the onset of chaos. Illustrated in Fig.2.15.

Poincaré section A Poincaré section, or first return map, is used to study the dynamics of a continuous flow by reducing it to a discrete map. This is achieved by constructing an $(n - 1)$-dimensional hyper-plane orthogonal to the system dynamics in n dimensions and then observing the intersection of the system trajectories and this plane. The principle is illustrated in Fig. 2.16.

Saddle A fixed point is a saddle if points nearby to it may both converge and diverge from it (mathematically, this is characterised by both positive and negative eigenvalues of the linearisation of the map). Some directions are stable and some are unstable. Nonetheless, unless the system state happens to lie exactly on a stable direction (no unstable component), then the system state will *eventually* diverge from the fixed point.

Stable A fixed point is stable if points nearby to it converge to it (mathematically; this is characterised by negative eigenvalues of the linearisation of the map).

Unstable A fixed point is unstable if points nearby to it diverge from it (mathematically, this is characterised by positive eigenvalues of the linearisation of the map).

Exercises

1. Define the n-th term in the Fibonacci sequence $f(n)$ by $f(n) = f(n-1) + f(n-2)$. Does $f(n)$ grow faster, or slower, than exponential?

2. Prove by induction the explicit form of the Fibonacci sequence. That is, suppose that $f_t = \frac{1}{\sqrt{5}}(\phi^t - (1-\phi)^t)$ for some specific value of t (not for all t, yet). Then show that this implies that it is true for $t+1$. Finally, confirm that it is true for the particular value $t = 1$. Hence, one can deduce that it is true for $t = 2, 3, 4, \ldots$.

3. The Fibonacci model of population growth dictates that the population on day n, f_n, is given by

$$f_n = f_{n-1} + f_{n-2}.$$

   (a) Convince yourself that this coincides with the description given in the text.

   (b) Write this difference equation as a two-dimensional matrix equation relating $\begin{bmatrix} f_{n+1} \\ f_n \end{bmatrix}$ to $\begin{bmatrix} f_{n-1} \\ f_{n-2} \end{bmatrix}$.

   (c) Solve this equation for $f_0 = f_1 = 1$ and hence show that Fibonacci's model leads to geometric growth in the rabbit population.

   (d) Describe how this is related to the differential equation model of exponential growth $\frac{df}{dt} = \lambda f$.

4. Solve $\frac{dN}{dt} = rN\left(1 - \frac{N}{K}\right)$. Hence, or otherwise, show that the solution has (in either case) one stable and one unstable fixed point, and describe their stability.

5. Let x_t denote the concentration (in parts per million, or ppm) of a certain hormone (triiodothyronine) in the bloodstream at time t (in hours). The hormone is consumed by various metabolic processes at a constant rate δ (ppm/hour). The thyroid gland regulates concentration of this hormone. If the hormone concentration at time t is below level c_0 (ppm), then the thyroid gland will produce triiodothyronine at a rate d_1 (ppm/hour). If the concentration is above the level c_0 (ppm), then production of triiodothyronine will be at a lower rate $d_2 < d_1$ (ppm/hour).

   (a) Write down a difference equation for this system.

   (b) Under what condition(s) will the hormone concentration remain stable (i.e. not increasing or decreasing indefinitely)?

   (c) What is the effect of the quantity $(d_2 - d_1)$? What happens when $(d_2 - d_1)$ is either large or small?

   (d) Draw a graph of the expected behaviour of this system under the physiologically normal state.

   Hyperthyroidism and hypothyroidism are two medical conditions corresponding to over- and under- active thyroid production of (various hormones including) triiodothyronine.

   (e) Explain how these two conditions may manifest in this model.

   (f) Suggest how (if at all) this model would need to be modified to provide a physiologically reasonable model of these conditions.

6. The model in Exercise 5 (a) is somewhat unrealistic as it is discrete in time and provides only a step change in the hormone production rate.

   (a) Suggest a continuous time model which captures the same physiological behaviour for the case when hormone production rate $d(x)$ is a continuous function of the current hormone level.

   (b) Sketch the expected shape of $d(x)$ so that the results are physiologically consistent with the situation in Exercise 5 (f).

7. Finally, suppose that the model in Exercise 6 (a) is to be generalised further so that the consumption of triiodothyronine is a continuous function $c(t)$ with a period of 24 hours $c(t + 24) = c(t)$. Sketch and describe the model behaviour $x(t)$.

8. (a) Consider the modified logistic map $x_{t+1} = rx_t(1 - x_t^3)$. Find the fixed points of this map and describe their stability.

   (b) Describe (in words) how you expect the behaviour of this system to differ from the standard logistic map.

9. In type 1 diabetes, sufferers lack insulin-producing beta cells. The lack of these beta cells leads to a lack of insulin and it is this insulin which should regulate the uptake of glucose from the blood. Hence, without sufficient insulin, excessive glucose will be present in the body which in turn leads to the clinical symptoms of diabetes. Let the hourly concentration of glucose be given by x_t and the effect of insulin on absorption of that glucose be determined by Ex_t, that is

$$x_{t+1} = \frac{bx_t^2}{1 + x_t^2} - Ex_t,$$

where the parameter $b > 0$ is the maximum glucose capacity.

(a) Sketch this map for $E = 0$, and hence describe the behaviour of $x_t \to 0$ and for $x_t \gg 0$.

(b) Keep $E = 0$; find the fixed points of the map and describe their stability.

(c) Describe in physiological terms the expected behaviour from this model in the absence of insulin-controlled glucose absorption.

(d) Suppose that the glucose capacity is 10 mM. If $E = 0$, express the steady-state concentration as a percentage of that capacity.

Chapter 3

Observability of Dynamic Variables

3.1 Bioelectric Phenomena Measurement

In the last two chapters, we introduced several interesting dynamical phenomena in various biological systems. We saw that the complex observed behaviour could be described in a variety of ways and that under certain situations, external changes can affect the behaviour of the system — even to the extent of altering the fundamental type of behaviour. In Chapter 2 the logistic map gave us a neat toy example of just this type of behaviour. But, one very important fundamental question remains: How do we know?

How do we know what the system is doing when we cannot measure exactly what type of behaviour is being exhibited. In certain, very simple cases, it is possible to measure all the relevant information: in the case of the rabbits, for example, it is possible to have a fairly complete picture of the population[1]. But, in general this is not enough. How do we know that by simply measuring the movement of the abdomen of a sleeping infant that one has enough information to describe the state of respiration. Moreover, how can one go about making measurements that are sufficient? In this chapter we will address both these questions: both the engineering challenges of measuring biophysical system parameters and the mathematical problems of how much measurement is enough.

Fortunately, the human body is an electrical machine. When muscles contract, they do so in response to electrical signals conducted along nerve cells. When the heart pumps — creating mechanical force — it does so as a result of muscle contraction, and the electrical activity that causes this is relatively easy to measure. The brain too is an electrical system (well, actually it is an electrochemical system with signals being transmitted both electrically as well as chemically — with the help of neurotransmitters). Within the brain the individual neuronal cells generate small electrical potentials (of the order of ± 70 mV) and the result of the massive orchestration of billions of these cells can then be measured as a changing electrical potential — either from the exterior surface of the head or by placing electrodes internally. Similarly, the

---

[1] If we leave aside sexual preference and proclivity of individual animals.

activities of all other muscles can be measured electronically. As we will see in Chapter 5.1, this is essentially a result of electrical activity — due to the movement of chemical ions carrying electrical charge across cell boundaries — in various cellular systems in the body.

In addition to measuring bioelectric phenomena directly (by measuring electrical potential), it is also quite common to measure the result of the bioelectric phenomena. Usually this means physical movement, and usually this can be measured with some simple form of displacement transducer — the effect of which is to convert the physical movement back into an electrical signal. In Section 3.2, we will consider in a little more detail the various different techniques for measuring a variety of bioelectric phenomena in which we are interested. In Section 3.6 we briefly discuss the vast topic of biomedical imaging. At the end of this chapter, in Section 3.7, we will return to the deeper mathematical problem of how much measurement is sufficient.

3.2   ECG, EEG, EMG, EOG and All That

The simplest form of bioelectric sampling we can consider is the direct measurement of electrical potential via electrodes. The electrode can be one of many forms — depending on the intended application. In Fig. 3.1 we depict just two basic types, and in Fig. 3.2 we have typical examples. The objective of each type of electrode is basically the same — to create an electrical circuit from an external voltmeter and the human body. Of course, human tissue conducts electricity fairly well. The internal electrical resistance of the human body is around 100–600 Ω. The main challenge is getting access to the internal workings of the human body. One approach is to use needle electrodes (pictured in Fig. 3.1) to penetrate the surface. However, for many applications (and most patients), a less invasive and inconvenient alternative is desirable. Unfortunately, the external resistance of the human body (that is, dry skin) is around 500 kΩ — one therefore needs to ensure a good electrical contact. Typically this is achieved with electrically conductive (i.e. electrolytic) gels within the electrode. These ensure a good electrical conduct between the metal electrode wire and the human skin, via the medium of the gel.

Once an electrical circuit (with known resistance) is established, it is then a simple matter to measure the potential difference (voltage), amplify and digitise as necessary. There are of course several applications of this basic technology. Of these, *electrocardiographs* (ECGs) are employed to measure the activity of the heart, *electroencephalographs* (EEGs) measure electrical activity in the brain, *electromyographs* (EMGs) measure electrical activity in various muscles, and *electrooculagraphs* (EOGs) measure eye movement. Of

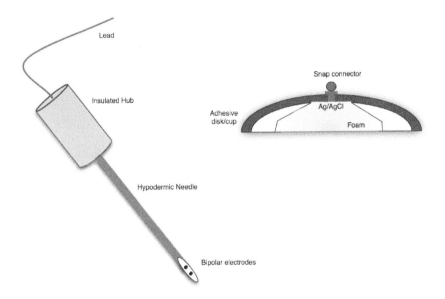

Figure 3.1:   **Electrodes.** Cartoon representation of the components of surface (on the right) and sub-dermal (left) electrodes. The sub-dermal (or needle) electrode consists of a hypodermic needle fitted with a bipolar electrode. The surface electrode is unipolar (multiple electrodes will be placed at different locations to create a circuit) and consists of a metal conducting snap connector fitted in an electrolytic (electrically conducting) gel. The electrolytic gel acts as a mediator between the metal conductor and the subject's skin surface and provides a sufficiently low electrical resistance.

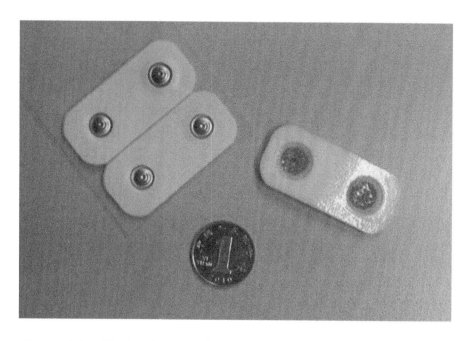

Figure 3.2: **Electrodes.** A photograph of standard ECG/EMG surface electrodes.

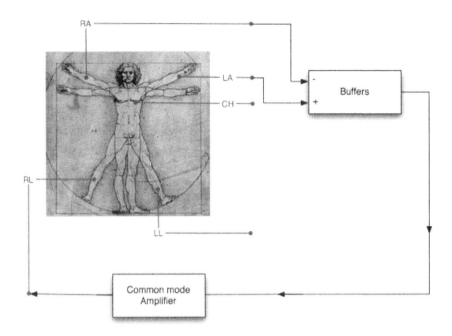

Figure 3.3: **ECG lead placement.** An illustration of typical electrocardiographic recording. Electrodes are placed on the extremities — at the wrists and ankle — as well as on the chest around the heart. Various combinations of leads are then compared to measure the voltage gradient in prescribed directions. In this case, by comparing the difference between the voltages at the left and right wrist to a ground at the right ankle, one can measure the voltage gradient horizontally across the chest. This configuration is known as *lead I.*

course, in each of these examples what one is essentially measuring is the electrical activity of different muscle groups. The key difference is only in the position and function of those muscles.

Take as a particular example the monitoring of the heart via ECG. The heart is the most electrically active muscle in the body. The electrical activity of the heart can be measured throughout the body. Even during vigorous exercise, ECG can be measured by electrical contact at the palms (this technology is routinely used in modern treadmills and running machines to monitor the heart rate of runners). Nonetheless, the heart is a three-dimensional solid mass of muscle and the electrical potential of the heart is a vector quantity — it has both a magnitude and a direction (in three dimensions). During the regular beating of the heart, both the magnitude and the orientation

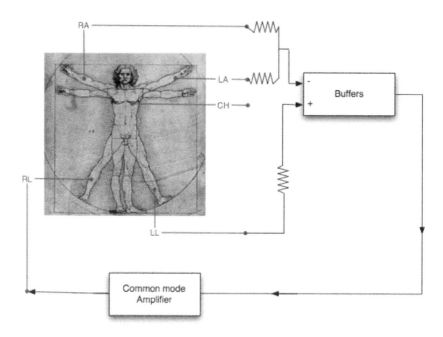

Figure 3.4: **ECG lead placement (lead VF).** In comparison to Fig. 3.3, this configuration compares the weighted average of voltage at the wrists to that at the left leg (again, using the right leg as a ground). With this arrangement (known as *lead VF*), one can measure the voltage change vertically through the trunk of the body.

of that beating will change periodically. Hence, to accurately measure the electrical state of the heart, one needs to obtain scalar measurements of the magnitude of the electrical activity of the heart in several different directions. Mathematically, the process is basically one of projecting the vector electrical potential v onto a particular measurement vector u (with $|u| = 1$) to obtain the scalar quantity $v \cdot u$. Physiologically, the process is systematised as the measurement of ECG with one of either three or twelve standard *leads* — two typical configurations are depicted in Figs. 3.3 and 3.4.

In the physiological usage, a lead is a particular channel in the multi-channel ECG recording, not a specific cable or wire. In the standard twelve lead ECG, for example, a patient will typically have separate cables connected to each of the four limbs and up to six locations on the left side of the chest (i.e. directly over the heart) — a total of ten wires and ten electrodes to deliver twelve leads. Each lead is composed of a specific combination of these cables (including a common ground point) and can measure electrical potential in a particular direction through the body. Hence, the measurement of the electrical signal in each of these twelve signals will result in a morphologically distinct scalar time series which can then be employed in combination to diagnose and study the electrical activity of the heart. We will return to the various nuances of electrocardiology in more detail in Chapter 4.

Electroencephalographic signals are similar — specific systems of electrode placement over the scalp (along with a common ground) exist and are widely used. These arrangements allow one to probe the individual electrical activity of different brain regions. But, of course these techniques are limited to regions on or near the surface of the brain and, partly because of the diffusive properties of the scalp, measure only relatively broad overlapping regions. We will have more to say about EEG signals in Chapter 4 when we discuss the purpose for and interpretation of clinical EEG recordings.

Similarly, EOG is an interesting specific application of the general methodology of EMG. By measuring the electrical activity of the muscles surrounding the eye, it is possible to detect which muscles are tense and therefore in which direction the eye is moving. Hence, by simply measuring the muscle activity in the facial muscles around the eye it is fairly easy to obtain an accurate picture of the subject's visual focus (i.e. what the person is looking at). Electromyography can of course be applied to a wide range of various muscle groups — often for either sports physiology or sleep study applications. In sleep studies, in addition to measure eye movement (which is used to diagnose dreaming states), EMG can be used to detect muscle relaxation (something which is indicative of deep as opposed to light sleep). In combination with ECG, EOG and often EEG, the EMG measurements can form a *polysomnographic* recording.

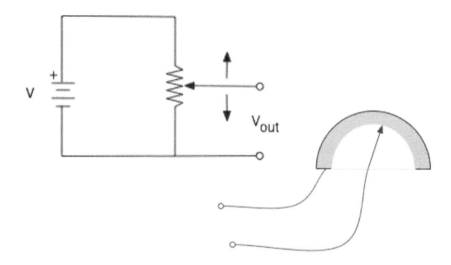

Figure 3.5: Potentiometer. A circuit realisation of a potentiometer (either linear or angular) is shown in the upper left part of the figure. As the position of the sliding pin (lower right side of the figure) changes, the observed voltage V_{out} changes in direct linear proportion to displacement. Similarly, the same circuit element is often realised as a physical angular displacement.

3.3   Measuring Movement

Many of the remaining (i.e. nonelectrical) physiological signals of interest can be observed by measured movement — either direct movement of the subject or movement induced through some other physical force. Displacement can be measured electrically by measuring change in one of the fundamental circuit variables, i.e. inductance, resistance or capacitance; or by measuring changes in electromotive force. Here we give brief examples of each. In each case, the basic principle which must be observed is the construction of a circuit element with some component variable by displacement. One can then observe the effect of any displacement on the behaviour of the circuit.

The simplest example of this is the standard potentiometer. Angular, or lateral, movement of a pin alters the length of resistance under load, and therefore, via Ohm's law,

$$V = IR, \tag{3.1}$$

the measure voltage. The principle is illustrated in Fig. 3.5. Suppose that the resistor illustrated has total resistance R and that a fraction a $(0 \leq a \leq 1)$ is beneath the pin. The observed output voltage V_{out} is given by $V_{\text{out}} = I(aR) = aV$ and is proportional to the position of the pin a.

Change in the resistance of a material is exploited at a more basic level by various types of strain gauges — often employed to measure change in fluid pressure. Recall that the resistance of a cylindrical rod of a conducting material R is proportional to the ratio of the length ℓ to the cross-sectional area A

$$R = \rho \frac{\ell}{A}, \tag{3.2}$$

where ρ is the material resistivity. This relationship is intuitive and completely analogous to the flow of water in a pipe — fatter pipes are good, longer pipes are bad. Now, suppose that a particular conductor (which we model as a stiff cylindrical rod) is placed under strain so that it is stretched $\ell \to \ell'$ and $A \to A'$. Of course, the total mass of material must be constant, $\ell A = \ell' A'$. Suppose that $\ell' = r\ell$; then $A' = \frac{1}{r}A$ and hence

$$R' = r^2 R \tag{3.3}$$
$$= \left(\frac{\ell'}{\ell}\right)^2 R.$$

The principle is illustrated in Fig. 3.6.

In a similar vein, one can exploit change in both capacitance and inductance with the physical variation in the structure of a capacitor or inductor. We will consider the case of an inductor first. An inductor is basically a conducting coil wrapped around an electrically permeable core. The important factor here is that the inductance L of a given inductor is related to voltage and current according to

$$V = L \frac{dI}{dt}. \tag{3.4}$$

The inductance itself depends on its physical structure

$$L = G\mu n^2, \tag{3.5}$$

where μ is the permeability of the medium (which may be expressed in terms of the permeability of free space μ_0 and a relative permeability μ_r as $\mu = \mu_0 \mu_r$) and n is the number of turns on the inductor. The conductance G is described as a "geometric form factor", what this means is that it depends on the shape of the inductor core. We can exploit the relation (3.5) in two separate ways: either by manipulating G or altering n. One way which G can be manipulated is if the inductor is constructed so that the core size changes. An application of this would be in measuring changing abdominal cross-sectional area (this is called inductance pleythsomnography when it is done as part of a sleep

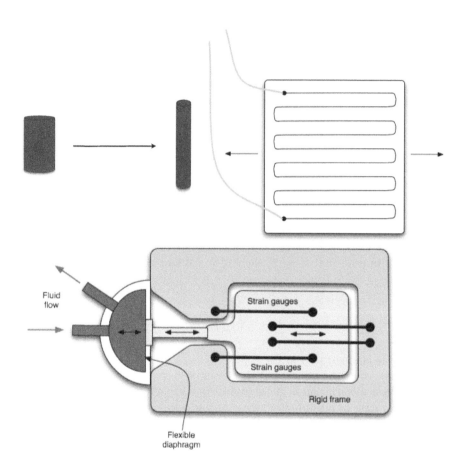

Figure 3.6: **Strain gauges.** Strain gauges exploiting changing resistivity in a metal conductor. As the conductor is elongated, its cross-sectional area is decreased. One can compound this effect in a bounded (upper right) or unbounded (lower) strain gauge. In a bounded strain gauge, a metal conductor is embedded (bonded) to a solid nonconducting substrate. As that substrate is put under strain, several lengths of conductor are elongated in a predictable fashion. In an unbonded strain gauge the central plate is free to move laterally, but that lateral movement causes sections on conductors to alternatively lengthen and shorten. Such an arrangement could be exploited (as shown here) to measure change in fluid flow pressure. The chamber on the right contains a fluid vessel with a flexible diaphragm connected to the moveable portion of the strain gauge. Increases or decreases in fluid pressure impinge on that diaphragm, causing the moveable central portion to move laterally, which affects the observed strain. See colour insert.

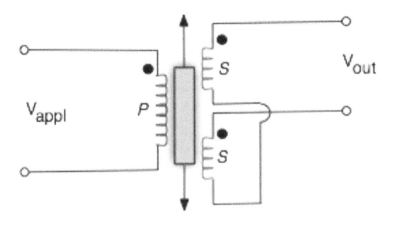

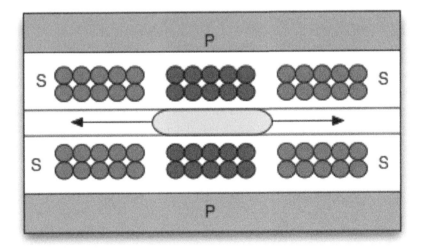

Figure 3.7: **Linear variable differential transformer.** An LVDT is illustrated both as a circuit diagram and a physical cross-section. Note that as the central core moves (vertically in the circuit, horizontally in the cross-section), the region of mutual inductance increases or decreases linearly. See colour insert.

study, commonly for infants, to measure respiratory effort). The wire coil (encased in an elastic band) is wrapped around the subject's abdomen. As the abdomen expands and contracts, the form factor G changes proportional to cross-sectional area and the effect is observed as a change in inductance. Conversely, one can construct a linear variable differential transformer — essentially a pair of couple inductors, with a moveable core. As the core of the LVDT moves, a greater or lesser proportion of the inductor is functional and the mutual inductance decreases in proportion to that change. The principle is indicated in Fig. 3.7.

Now, for the case of capacitance. A capacitor with capacitance C relates the voltage over it to the current through it according to

$$I = C\frac{dV}{dt} \qquad (3.6)$$

where the capacitance is constructed as a pair of charged plates of cross-sectional area A separated by a distance d. The capacitance of such a capacitor is given by

$$C = \varepsilon\frac{A}{d}, \qquad (3.7)$$

where ε is the dielectric constant of the medium separating the two plates (just as for inductors, the dielectric constant $\varepsilon = \varepsilon_0\varepsilon_r$ where ε_0 is the dielectric constant of free space and ε_r is the relative dielectric constant of the medium).

Finally, we conclude this section with a brief description of the Fleisch pneumotachometer. This device is used to measure air flow rate and does so by exploiting a pressure differential and measuring this pressure change with one of the displacement transducers described above. The basic principle is described in Fig 3.8.

3.4   Measuring Temperature

Temperature measurement, over a range of temperatures relevant for physiological processes, is a little different. While it is true that metallic conductors will change their properties in response to temperature changes, they do not typically do so over the range of temperatures that are physiologically relevant — say 35°C to 42°C. To measure such changes, one needs to rely on a special group of compressed sintered metal oxides. These are metal compounds (metal oxides) that are manufactured (sintered) and treated (compressed) under special conditions that cause them to exhibit a sensitive dependence on temperature. Unlike normal metals, which increase their resistance with temperature, the resistance of these thermistors decreases exponentially with

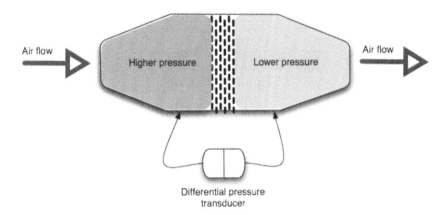

Figure 3.8: **Fleish pneumotachometer.** As air flow (from left to right) encounters a fine grating in the central region of the pneumotachometer, a pressure gradient proportional to the flow rate will occur. A displacement transducer (for example, either the LVDT or the unbonded strain gauge described above) can be used to measure this pressure differential and thereby observe the flow rate.

temperature

$$\frac{R_1}{R_2} = \exp\left[\beta\left(\frac{1}{T_1} - \frac{1}{T_2}\right)\right], \qquad (3.8)$$

where R_1 is the resistance at temperature T_1 and R_2 is the resistance at temperature T_2 (measured in Kelvin). The constant $\beta > 0$ depends on the physical properties of the particular thermistor.

3.5 Measuring Oxygen Concentration

There is one more scalar physiological parameter which is regularly measured and which we will briefly consider here — blood oxygen concentration. Blood that is rich in oxygen is redder; deoxygenated blood is bluer. In fact, the amount of oxygen in the blood can be directly related to the quantity of red light of a particular wavelength which is absorbed by the blood. Hence to measure blood oxygen concentration, one shines red light of two similar wavelengths through a body extremity (depending on the size/age of the subject, a finger tip, ear lobe or foot will usually work well). According to Beer–

TABLE 3.1: Typical acoustic impedances.

Material	c (ms^{-1})	ρ (kgm^{-3})	Z (MegaRayls or kgm^{-2}s^{-1})
Air	340	1.2	0.0004
Bone	3360	1790	6.00
Blood	1550	1040	1.61
Fat	1450	950	1.38
Liver	1570	1050	1.65
Muscle	1580	1040	1.645
Water	1480	1000	1.480

Lambert's law, the transmitted (output) light intensity P_t is related to the incident (input) light intensity P_i by

$$P_t = P_i 10^{-abc}, \tag{3.9}$$

where a is a constant that varies with the wavelenth (hence the need for two similar wavelengths), b is the thickness through the sample (which will also be the same for both samples) and c is the dependence on the absorption of that wavelength. That is, the wavelength corresponding to oxygenated blood c will be proportional to the concentration of oxygen in the blood.

3.6   Biomedical Imaging

Before we conclude this chapter on how and how much to measure, it is both instructive and interesting to briefly consider the implementation of various medical imaging technologies. The instrumentation techniques described earlier in this chapter, and indeed the focus of this book, is on scalar time series measurements. Nonetheless, several techniques exist for constructing images of the internal structure of the human body. In this section we focus on two that are conceptually simple: ultrasound and x-ray.

In Fig. 3.9 we illustrate an ultrasound image of a 4-month-old foetus *in utero*. The image is produced by bouncing sound waves from the exterior of the mothers abdomen and looking for reflections (or echoes) which will correspond to changes in the acoustic properties of the medium. By observing the timing and extent of these reflections, it is possible to determine the relative acoustic density of various materials and the distance from the source transducer. Figure 3.10 provides a cartoon of the general procedure.

The acoustic properties of each material are measured by its *acoustic impedance* Z which is the product of the material density ρ and the speed of sound in that material c

$$Z = \rho c. \tag{3.10}$$

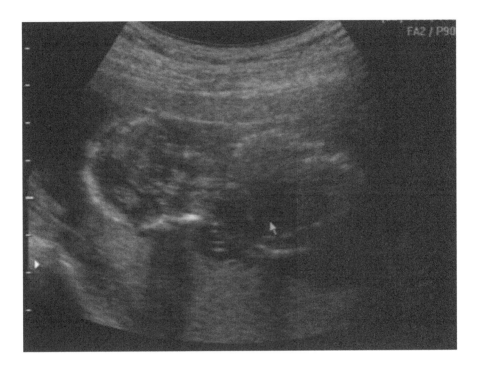

Figure 3.9: **An ultrasound image.** The head is on the left, the spine on the top right, the arrow marks the approximate location of the heart (which can be seen to beat in real time). The shell of the developing cranium is visible on the left-hand side of the image, the face (the outline of the lower jaw is visble) is downwards. Just below this, the digits of the hands are visible and next to this are the feet, curled under the spine (the vertebrae are visible on the top right).

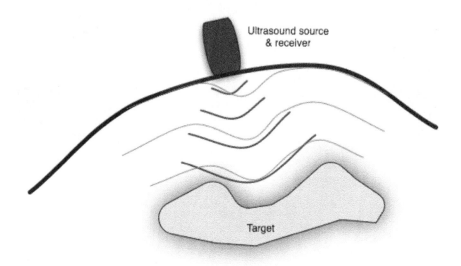

Figure 3.10: The basic principle of ultrasound. The ultrasound transceiver — both a transmitter and a receiver — transmits a series of ultrasound pulses into the tissue. As these pulses travel through the tissue mass, reflections (echoes) will be generated at the boundary between materials of differing acoustic density. These reflected pulses are returned to the transducer. The strength of the reflection can be used to determine the variation in acoustic density; the timing between transmission and reception of the corresponding pulse can be used to determine the relative acoustic density and the distance/thickness involved.

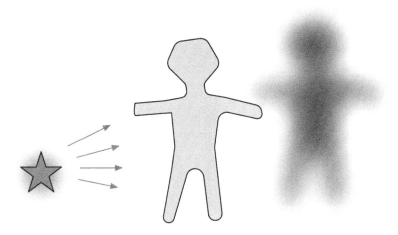

Figure 3.11: X-ray imaging. The basic principle of x-ray imaging is to use a radiation source to cast ionising radiation over the subject. The degree to which that radiation is absorbed by the subject manifests as a darker radiation shadow behind the subject.

At the sharp boundary between materials with acoustic impedance Z_1 and Z_2, the reflection factor is

$$R_f = \frac{Z_2 - Z_1}{Z_2 + Z_1}. \tag{3.11}$$

Table 3.1 lists the acoustic impedances of several common biophysical materials.

Of course, to obtain a two-dimensional image, one needs to make measurements in different directions. This can be seen in the example image in Fig. 3.9. In this image one can see the semicircular geometry of differing sound pulse propagating from the single transducer. X-ray works with similar principles. However, unlike ultrasound imaging , x-rays rely on ionising radiation and exposure to it must be severely limited. Essentially, a radiation source is placed on one side of a target image and a radiation detector (this could be either a photographic plate or an electronic detector) is put on the other side of the target. The amount of radiation incident on the radiation detector will be reduced due to radiation absorbed by the target. Regions of great density will absorb more radiation. In this way one obtains an image which is essentially a shadow of the material density of the target — see Fig. 3.11.

Essentially each x-ray measurement made by the radiation detector can only measure the material density of one path through the target. In most

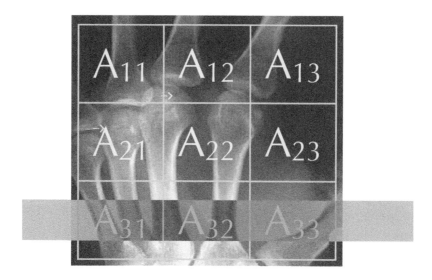

Figure 3.12: **Reconstruction of a solid image from x-ray observations.** In this example the image consists of only 9 pixels, A_{11} to A_{33}, but we wish to determine numerical values for each. However, the individual pixels cannot be accessed directly; rather, an x-ray beam can be shone through the sample and the contribution of the ensemble of pixels is observed. In this example, by observing a horizontal slice from left to right, we can measure a value for $A_{31} + A_{32} + A_{33}$.

cases, what is needed is a complete picture of the material density at every point along the path, and this procedure must be repeated for an ensemble of distinct paths. Essentially this problem is the foundation of a large area of mathematics called *inverse problems*. The usual direction is to start from a cause and determine the effect: that is, for a given x-ray beam and a particular target object, what will be the shadow cast? Inverse problems work in the reverse; given the solution (a particular shadow profile), what question (what object) could have generated it? The principle is reminiscent of the 1960s American television game show, *Jeopardy!* in which contestants attempt to formulate the (correct) question which would generate a particular answer.

The mathematical solution to generate the cross-sectional profile of a target from a series of x-ray shadows is a little more complicated. In its complete form, the Radon transform is beyond the scope of this presentation. Instead we will present a conceptually easier brute-force approach. Suppose that the image we wish to reconstruct consists of only nine pixels in a 3×3 grid and that the imaging process via x-ray shadow observation is essentially equivalent to the sum of the components on any axis. Figure 3.12 shows one such observation. Now consider various row-wise and column-wise operations to observe horizontal and vertical slices through the matrix

$$\begin{bmatrix} A_{11} & A_{12} & A_{13} \\ A_{21} & A_{22} & A_{23} \\ A_{31} & A_{32} & A_{33} \end{bmatrix}.$$

One can obtain measurements for the shadow intensities R_1 to R_6 that satisfy the following six equations

$$A_{11} + A_{12} + A_{13} = R_1$$
$$A_{21} + A_{22} + A_{23} = R_2$$
$$A_{31} + A_{32} + A_{33} = R_3$$
$$A_{11} + A_{21} + A_{31} = R_4$$
$$A_{12} + A_{22} + A_{32} = R_5$$
$$A_{13} + A_{23} + A_{33} = R_6.$$

Of course, to solve for nine unknown variables, this is not enough information. However, we may continue in the same way and take diagonal slices

$$A_{21} + A_{11} + A_{12} = R_7$$
$$A_{31} + A_{22} + A_{13} = R_8$$
$$A_{32} + A_{33} + A_{23} = R_9$$

which gives us a total of nine equations in nine unknowns. Unfortunately, these equations are not independent, and we continue

$$A_{12} + A_{13} + A_{23} = R_{10}$$
$$A_{11} + A_{22} + A_{33} = R_{11}$$
$$A_{21} + A_{31} + A_{32} = R_{12}$$

which finally provides sufficient information to solve for the nine unknowns, via matrix inversion. Of course, this is only a toy example, and the method will become unmanageable for larger sets of observations and higher image resolution. Hence, the Radon transform can be used to systematise and implement the process for larger systems of equations and provide a computational solution to this particular inverse problem.

3.7   The Importance of Measurement

While it might seem that the topic of this chapter really belongs in a much larger text on biomedical engineering instrumentation, this is not really our

point. The main focus of this text is dynamical behaviour of biological systems. In all cases, the core of this topic is: just what is it that can be
measured? Without measurement of the underlying system, it is futile to try
to describe the system dynamics. Moreover, as we will see in the next chapter,
when measurement is combined with an understanding of the dynamics which
are expected, it is possible to provide powerful diagnostic information.

Nonetheless, what needs to be measured? How much information do we really need? With the exception of the imaging techniques described in Section
3.6, all the techniques described in this section measure single biological variables: either a voltage, or temperature or physical position of a single element
of the body. It turns out, thanks to some rather fancy mathematics, that in
general, this is sufficient. For either differential or difference equation systems
(equations with a time delay are a little more complicated, but one can still
achieve the same result), the current state of the system can be described by
a series of n variables. For example, with a system of n differential equations, the value of those n variables (or their first derivatives) is enough to
know what's going on in the system. But, a system of n first-order differential
equations can always be re-written as a single n-th-order system. Hence all
we really need to know is the value of one variable and $(n-1)$ derivatives.
However, to numerically estimate these derivatives, it is sufficient to measure
at different times and look at how the value changes. Hence, n successive
measurements of a single variable are sufficient. We will return to this idea in
more detail in Section 4.4.

Of course, in some applications, what is sufficient is not what we use. Medical imaging presents a very good example. ECG and EEG are also often
done with multiple measurements, the basic reason for this is two-fold: first,
because we can (at no great additional cost) and second, because this is what
the medical practitioners are used to interpreting. While a single scalar ECG
is sufficient, measurement of the multiple different leads provides the cardiologist with a more convenient visual interpretation of the dynamical behaviour
of the heart. In the next chapter we will return to this particular example
and show how these different methods are, nonetheless, equivalent.

Finally, there is a second sort of measurement which is important for the
purpose of system identification. Very often we are able to write down a set of
differential or difference equations which we suppose give an accurate description of what is really going on. But these equations will have parameters, the
values of which are not generally known. In Section 2.4 we had an example
of this. Equations (2.8) and (2.9) gave an accurate description of system dynamics, but depended on multiple parameters. In the case of Section 2.4, as
well as many other typical examples, it is possible to estimate values for these
parameters by observing what the system does. For example, in the case
of prostate cancer, all the parameters related to different cell growth rates
and could therefore easily be estimated from observing the changing system
behaviour.

3.8 Summary

The human body functions because of muscular activity. That muscular activity in turn is driven by electrical impulses. By harnessing various measurement techniques to observe these electrical impulses, it is possible to provide a picture of the dynamical activity of the various physiological component systems in the body. Electrocardiograms, electroencephalograms, electrooculagrams and electromyograms are all specific examples of such measurements. In some cases one can choose to measure physical displacement instead: either by measuring change in resistance, change in capacitance or change in inductance in various instruments. Nonetheless, we have seen that in all these situations it is possible to measure the dynamical state of the system merely by observing the changing behaviour of one variable of that system.

Glossary

Electrocardiography The observation of the dynamics of the heart by measuring changing electrical potential.

Electroencephalography The observation of activity of the brian by measuring changing electrical potential.

Electromyography The observation of changing muscular activity by measuring changing electrical potential.

Electrooculagraphy The observation of movement of the eye by measuring changing electrical potential of the muscles surrounding the eye.

Electrophysiology The observation of physiological phenomena by measuring changing electrical potential.

Exercises

1. (a) A regular copper conducter (resistivity 16.8 nΩm) with circular cross-section is deformed to an oval cross-section with minor axis twice the major and stretched to three times its original length. If the original resistance of this 20-cm-long wire was 120 mΩ, what is the value of resistance after the deformation described above?

(b) Silicone rubber has a dielectric constant of 3.2 and is used to form a compressive strain gauge with two plates of area 1.3×10^{-2} m^2. If the distances between the plates at rest is 15 mm and this is decreased to 10 mm, what is the observed change in capacitance?

(c) A copper oxide thermistor is observed to have a resistance of 150 Ω at 5°C and 15 Ω at 100°C. At what temperature will this thermistor exhibit a resistance of 65 Ω?

(d) Compute the reflection factor at the boundary between muscle and each of air, bone, blood, fat and liver tissue.

(e) Hence, an ultrasound pulse reflection (reflection factor 0.57) is observed after travelling for 15 µsec. through an unknown material. How thick is this unknown material; and hence, what two materials and in which order were involved in this reflection?

2. The inductance of a solenoid is given by $L = \frac{\mu_0 \mu_r n^2 A}{\ell}$, where μ_0 and μ_r are the permeability of free space and the relative permeability of the core; n is the number of turns; A is the cross-sectional area; and ℓ the length of the coil. Determine an expression for the "geometric shape constant" given above. Hence, show that inductance plethysmography results in change in inductance proportional to change in cross-sectional area.

3. Show that the change in length $\Delta \ell$ of a strain gauge results in a change in resistance ΔR proportional to $\frac{2 \Delta \ell}{A}$ (approximately).

4. Describe a circuit which may be used to realise a linear displacement transducer utilising a differential (three parallel plate) capacitor.

5. (a) Compute the inductance in a steel core inductor with geometric shape constant 2×10^{-6}. The permeability of steel is 875 µN/A^2.

(b) What is the change in resistance of a tungsten strain gauge (resistivity of tungsten is $\rho = 5.6 \times 10^{-8}$ Ωm) if the length is doubled?

(c) What is the capacitance of an open air capacitor $\epsilon_r \approx 1.00054 \approx 1$, $\epsilon_0 = 8.8541878 \times 10^{-12}$ F/m) consisting of two plates 10 mm-square separated by 1 mm?

(d) If the resistance of a zinc-oxide thermistor at 0°C is 100 Ω, what is the resistance at 30°C? (Take $\beta = 3500$K.)

(e) Compute the change in an inductance core if the core is deformed such that its length is doubled and its cross-sectional area is correspondingly shrunk (in such a way that volume is preserved).

(f) What is the change in resistance of a steel strain gauge if its length varies by 10%?

(g) What is the capacitance of a silicon-filled three-plate capacitor if the plates are 10 mm^2, the outer two plates are 3 mm apart and the middle plate is located mid-way between the other two? Suppose the middle plate moves 0.2 mm; what is the change in capacitance?

(h) Estimate the material constant β for a copper-oxide thermistor with the following measured temperature-resistance profile (for convenience, take $T_0 = 0°C$).

temperature (°C)	−10	0	20	100
resistance (Ω)	140	100	60	14

(i) If the concentration of O_2 in the blood increases from 90% to 99%, how does the incident light intensity of a red-light oximeter change?

6. Beer–Lambert's law states that incident light intensity P_t is related to a wavelength constant a, distance b and target chemical concentration c. Suppose that two similar wavelengths (same a) are passed through a sample (finger). One wavelength is sensitive to the concentration of O_2 in the blood C_1, the other is not $C_2 = 1$. Deduce a relationship for C_1 in terms of the observed incident light intensities $P_{t,1}$ and $P_{t,2}$ at the two wavelengths.

7. (a) A linear variable displacement transformer is mounted with a spring in compression (remember that the force/displacement relationship of a spring is given by Hooke's law). The apparatus is then connected to a closed face mask at a constant temperature. Derive a relationship between pressure in the mask and the measured voltage. You should clearly state both the nature of the relationship and also which constants are involved.

(b) For the purposes of usefully measuring body temperature, a thermistor must be sufficiently sensitive to differentiate between a change in temperature of 0.1°C at normal body temperature. If resistance can be determined to an accuracy of ±0.1%, what is the required material constant? Would a metal oxide with a material constant of 1000 be sufficient?

(c) A reflection factor of 0.75 is observed at the boundary of two materials. What is the relative magnitude (i.e. $\frac{Z_2}{Z_1}$) of the two materials? If the first (less dense) material is blood, what is the acoustic impedance of the second material?

8. A linear variable differential transformer (LVDT) with a circular permalloy core ($\mu = 8 \times 10^{-3}$ N/A^2) of diameter 5 mm is used in the construction of a pressure transducer for a Fleish pneumotachometer. The LVDT is placed between the two chambers of the pneumotachometer so that the displacement from a neutral centre position, d, of the transformer

core is proportional to the difference in pressure at inlet P_{in} and outlet P_{out}. That is, $d = k|P_{in} - P_{out}|$. In this particular pneumotachometer, the rate of flow of air v is linearly proportional to the pressure difference between inlet and outlet chamber via the resistance of the mesh screen R.

The *vital capacity* of a healthy adult is the maximum amount of air which can physically be exhaled and is typically about 4.6 L. During a calibration test for this device, this volume is forcibly exhaled over a period of 3 sec. Over that 3-second period, the average measured output of the LVDT is 5.4 mV, corresponding to a displacement of the LVDT core of 15 mm.

(a) Derive the relationship between air flow and voltage.

(b) What is the device constant $\frac{R}{k}$? Give appropriate units.

(c) What will the measured voltage be when the air flow is 200 mL/sec.?

(d) What do you expect the measured voltage to be at rest?

9. An ultrasound pulse emitted at 30 kHz generates two reflected waves with delays of 82 μsec. and 133 μsec and reflection factors of 0.088 and 0.57. What three materials are involved, in what order, and with what thickness? (Assume that only the materials described table 3.1 may be involved.)

10. Consider the simple CAT scan example given in lectures. Suppose one wishes to obtain two-dimensional measurements of the 9 pixels in a 3×3 pixel array. A sensor can make three measurements horizontally and three measurements vertically across the array. A further three measurements can be made diagonally (one measurement taking in the main diagonal, and another two taking in distinct groups of three off-diagonal elements). That is, suppose that the two-dimensional slice to be imaged can be represented by the matrix

$$\begin{bmatrix} A_{11} & A_{12} & A_{13} \\ A_{21} & A_{22} & A_{23} \\ A_{31} & A_{32} & A_{33} \end{bmatrix}$$

and that the nine measurements give nine distinct equations as follows,

$$R_1 = A_{11} + A_{21} + A_{31}$$
$$R_2 = A_{12} + A_{22} + A_{32}$$
$$R_3 = A_{13} + A_{23} + A_{33}$$
$$R_4 = A_{11} + A_{12} + A_{13}$$
$$R_5 = A_{21} + A_{22} + A_{23}$$
$$R_6 = A_{31} + A_{32} + A_{33}$$
$$R_7 = A_{12} + A_{13} + A_{23}$$
$$R_8 = A_{11} + A_{22} + A_{33}$$
$$R_9 = A_{21} + A_{31} + A_{32}$$

(the derivation of these equations is explained in lectures). Formulate this system as a matrix equation of the form $Ax = b$ (i.e. explain the role of A, x and b), and hence (or otherwise) show that it cannot be solved. Nonetheless, the system can be extended to one that is solvable by taking three further measurements in a different diagonal direction

$$R_{10} = A_{11} + A_{12} + A_{21}$$
$$R_{11} = A_{13} + A_{22} + A_{31}$$
$$R_{12} = A_{23} + A_{32} + A_{33}.$$

Show that this new extended system is solvable and then solve for the specific case where the twelve measurements are

$$\begin{bmatrix} R_1 \\ \vdots \\ R_{12} \end{bmatrix} = \begin{bmatrix} 2 \\ 2 \\ 2 \\ 1 \\ 2 \\ 3 \\ 2 \\ 1 \\ 3 \\ 2 \\ 1 \\ 3 \end{bmatrix}.$$

11. (a) What change in transmitted light intensity is induced by a reduction in blood O_2 concentration from saturation to 90% of saturation? 80%? 60%?

   (b) Quantify the change in resistance if a resistive strain gauge is lengthened uniformly by 40%.

(c) Characterise the observed reflection factors and time delay for an ultrasound pulse passing through 3 cm of fat followed by 2 cm of muscle before being incident on bone.

12. A linear variable displacement transducer is used to measure changes in air pressure during exhalation. The displacement of the transducer is found to be directly proportional to change in pressure within a sealed face mask. Suppose that the ratio of applied (input) to output voltage at rest is $\frac{v_{out}}{v_{appl}}$. When the applied voltage remains unchanged and the pressure within the mask increases by 20%, the output voltage v_{out} is observed to increase by 20 mV. What will the ouput voltage be, compared to v_{out}, if the pressure within the mask is 50% higher than the rest state?

13. Suppose the number of turns in the inductor used in the previous exercise is doubled. How will the inductance change?

14. If the resistance of a zinc oxide thermistor at 0°C is 400 Ω, what is the resistance at 38.5°C?

15. If the resistance of a thin metal wire is doubled under strain, how has the length changed?

16. Compute the index of refraction and the relative delays (time) involved in a muscle-bone transition (transition from muscle to bone) 2 cm from an ultrasound sensor.

17. (a) If a particular thermistor exhibits a resistance of 100 Ω at 40°C and 140 Ω at 20°C, what is the resistance at 60°C?

(b) Suppose an ultrasound signal is being used to image a sample consisting of layers of water, fat and bone (in that order, but of unknown thickness). Compute the reflection factor R_f for a transition from water to fat.

(c) Refer to part (b), above. If ultrasound reflection (for the water-fat and fat-bone transition, respectively) occurs after a time delay of 30 μsec. and 45 μsec, what thickness of fat and bone are involved?

(d) A differential capacitor is constructed from three plates 10^{-4} m² in area, separated (at rest) by gaps of 5 mm. Suppose the relative dielectric constant of the electrolyte involved is 2.6. What is the capacitance between just two of these plates?

(e) Refer to part (d), above. What changes (both increase and decrease) in capacitance will result from a 0.4 mm deviation of the central plate?

18. (a) What is the capacitance between two parallel plates, separated by an electrolytic solution of ethylene glycol ($\epsilon_r = 37$), if the area of each plate is 0.0045 m² and the distance between them is 5 mm?

(b) A temperature transducer is to be constructed from a specific sintered silver oxide compound which is observed to have the following temperature/resistance characteristic:

temperature (°C)	0	20	37
resistance (Ω-cm)	500	x	99

Compute the missing values.

(c) Compute the measured electrical resistance in a bonded strain gauge if a tungsten filament (resistivity 5.60×10^{-8}) of length 150 mm and diameter 0.8 mm is stretched to 120% of this length.

(d) The iPod/iPhone application iPulse[2] uses the devices camera (and LED flash) to measure pulse. Describe the biophysical principles on which such measurements could be made.

(e) Refer to the table in part(b). Why is it not possible to use such a device to measure blood oxygen saturation? What additional information (or hardware) would be required to do this?

---

[2]Not its real name.

Chapter 4

Biomedical Signal Processing

4.1 Automatic Analysis of Electroencephalograms

How can knowledge of the deterministic dynamics of a biological system help the medical practitioner? A very simple example of the application of a most basic level of knowledge is the analysis of electroencephalographic (EEG) signals. EEG recordings are made for a variety of clinical reasons, and by doing so, they provide a coarse measure of the mental state of the subject. The main weakness of EEG measurement, and particularly surface electrode EEG, is that such measurement can only provide a crude measurement of the actual activity of the brain. The measured surface EEG is a spatial average of the electrical activity of a vast number of individual neurons (which we will come to in a later chapter) diffused through cranial fluid, skull and skin.

Nonetheless, such an imprecise measurement can still be useful. Since the very earliest experiments with measuring EEG, it has been observed that the dominant frequency of electrical activity is associated with specific neural functions. The typical EEG signal is (rather arbitrarily) divided into five separate frequency bands δ (0.5–4 Hz), θ (4–8 Hz), α (8–13 Hz), β (13–22 Hz), γ (22–40 Hz). Of course, the actual division is not so clear-cut. Moreover, different sources will indicate slightly different frequency bands. Nonetheless, the definition of α to δ rhythm based on these frequencies is sufficient for us. Incidentally, the names α, β, γ and so on are assigned to different frequency bands based on the order in which they were first described. It does not, of course, coincide with the actual numerical bandwidths. Finally, we should not think that any activity outside these bands, below 0.5 Hz or above 40 Hz, can also be included in this definition. The numerical values here are just guidelines.

While this scheme provides nothing more than an approximate classification of activity of different frequencies, it is widely held that certain frequency bands are most commonly associated with certain types of activity. Broadly speaking, the five frequency bands are generally described in the following terms.

Delta δ (0.5–4 Hz): Associated most strongly with deep sleep (what is called stage 3-4 sleep).

Theta θ (4–8 Hz): Associated with lighter sleep (stage 2) and drowsiness (stage 1). More common in adolescence and early adulthood.

Alpha α (8–13 Hz): Present when awake but without sensory focus (i.e. eyes closed). Diminishes when eyes are open or when the subject's attention is focused. Characteristic of a relaxed alert state.

Beta β (13–22 Hz): Associated with active concentration.

Gamma γ (22–40 Hz): Associated with higher mental function, attentiveness or sensory stimulation.

It is clear that the higher frequency rhythms are associated with more active (more alert or concentrated) mental states. Nonetheless, the distinction between different rhythms is not so stark. Despite this, in the field of sleep studies, for example, there is a clear clinical interpretation as different levels of sleep are associated with different rhythmic states. Stage 1 sleep is associated with the drowsy state during onset of sleep. Once an individual is properly asleep, the sleep state cycles through stages 2, 3 and 4[1] and REM (or Rapid Eye Movement) sleep. The distinction between these various states can be diagnosed algorithmically by looking for changes in the dominant EEG rhythm during sleep. REM sleep occurs during dreaming and is often diagnosed with the assistance of electrooculagraphs (EOGs) and electromyographs (EMGs). During REM sleep the eyes move in a synchronous fashion (which can be diagnosed with EOG) and muscle tone is relaxed (which can be detected by EMG) — largely to prevent bodily movement during dreaming.

Let us now suppose that our simple aim is to detect *changes* in sleep state (or in some other state of mental function). The sleep states themselves are discrete and fairly well defined (the boundary between 3 and 4 is a little more tricky — and hence that boundary is now usually ignored) and hence they correspond to discrete states of mental function. The obvious "engineering" approach to solve this problem would be to bandpass filter the signal into the different frequency bands and look for which is dominant. Unfortunately, this is not a good strategy for two reasons: first, the boundaries between the different states, while definite, are also arbitrary and may not work equally well in all situations; and, second, this approach is (relatively) computationally expensive. It turns out that a generic alternative will serve us better.

The basic idea is the following. Since we are interested in how the dynamical structure of the system *changes*, we should be a *model* of that structure. The model should be tuned to the specific observed data. Then, we continue to test this model on new data as it comes in. When the system structure changes, the model will cease to work well (this is essentially what we mean by the structure changing). Hence, we construct a numerical measure of model

---

[1]As of 2004, stage 3 and 4 have been combined in standardised sleep scoring into a single state: stage 3.

performance and evaluate it on the incoming data. When the model performs worse than some pre-ordained threshold, then we can declare that the system has now entered a new state. At this point we start again.

1. Build a model f on the current data $x_{t_0-w}, \ldots, x_{t_0}$ (t_0 is the current time index and w is some window size).

2. Let $t_1 = t_0$.

3. Evaluate the model performance ξt_1 on the data $x_{t_1-w}, \ldots, x_{t_1}$. The simplest way to do this is to compute the model prediction error on that data window — that is, how well does the model describe each observation in that data window?

4. If $\xi_{t_1} \leq \xi_{t_0}$ or $\xi_{t_1} \approx \xi_{t_0}$, then increment t_1 and repeat step 3.

5. Since $\xi_i \gg \xi_{t_0}$, flag t_1 as the time at which the system state changes. Reset $t_0 = t_1$ and continue from 1.

The algorithm relies on the computation of model performance. For each window of length w we need to compute how well the model is doing. The most obvious way to do this is to test the model prediction performance. Essentially, the model f provides an estimate of x_{t+1} from the preceding points:

$$x_t \approx f(x_{t-1}, x_{t-2}, x_{t-3}, \ldots) \tag{4.1}$$

and hence the model performance can be computed as the average accuracy of these predictions

$$\xi_{t_0} = \frac{1}{w} \sum_{t=t_0-w+1}^{t_0} (x_t - f((x_{t-1}t, x_{t-2}, x_{t-3}, \ldots))^2. \tag{4.2}$$

So, now, the main question is what model f should we use, and how do we build it?

To answer this question, we do not completely abandon the engineer and the frequency domain ideas which are so obviously a feature of this particular system. Nonetheless, it turns out that it is far better to employ these methods in the time domain, and we do so with linear models.

The simplest time domain linear model is the autoregressive process of order 1, the AR(1) process:

$$\hat{x}_t = a x_{t-1} \tag{4.3}$$
$$e_t = x_t - a x_{t-1}.$$

That is, the next value x_t is simply the previous value x_{t-1} multiplied by some scaling a. Essentially, the model is just a. One chooses the value of a

which makes the average model error e_t as small as possible. Since the average model error is given by

$$\xi = E(x_t - ax_{t-1})^2, \tag{4.4}$$

we need to choose the fixed value of a to minimise ξ.

Of course, the solution is obvious: just as with ordinary functions of a single variable, to find the minimum (or maximum) we need to find when the derivative is 0:

$$\frac{\partial}{\partial a}[\xi] = \frac{\partial}{\partial a}\left[E\left((x_t - ax_{t-1})^2\right)\right]$$

$$= 2E\left((x_t - ax_{t-1})\frac{\partial}{\partial a}(x_t - ax_{t-1})\right)$$

$$= 2E\left((x_t - ax_{t-1})(-x_{t-1})\right)$$

$$= 0.$$

But, recall that $E(\cdot)$ is the expectation operator (that is, it returns the average of what's inside it) and that the autocorrelation $\gamma_{xx}(\tau)$ is defined to be

$$\gamma_{xx}(\tau) = E(x(t)x(t+\tau))$$
$$= E(x(t)x(t-\tau)). \tag{4.5}$$

Hence, $2E\left((x_t - ax_{t-1})(-x_{t-1})\right) = 0$ implies that

$$a = \frac{\gamma_{xx}(1)}{\gamma_{xx}(0)} \tag{4.6}$$

$$\xi = (1 - a^2)\gamma_{xx}(0).$$

That is, to determine the best model, we just need to compute the autocorrelation for $\tau = 0$ and 1.

Of course, this model is rather simple. In general, AR(1) will not be sufficient, so we extend the same basic idea to AR(2):

$$\hat{x}_t = a_1 x_{t-1} + a_2 x_{t-2} \tag{4.7}$$

$$e_t = x_t - (a_1 x_{t-1} + a_2 x_{t-2})$$

$$\xi = E(x_t - a_1 x_{t-1} - a_2 x_{t-2})^2$$

and seek the solution of

$$\frac{\partial}{\partial a_1}[\xi] = 0$$

$$\frac{\partial}{\partial a_1}[\xi] = 0,$$

which yields

$$\begin{bmatrix} \xi \\ 0 \\ 0 \end{bmatrix} = \begin{bmatrix} \gamma_{xx}(0) & \gamma_{xx}(1) & \gamma_{xx}(2) \\ \gamma_{xx}(1) & \gamma_{xx}(0) & \gamma_{xx}(1) \\ \gamma_{xx}(2) & \gamma_{xx}(1) & \gamma_{xx}(0) \end{bmatrix} \cdot \begin{bmatrix} 1 \\ -a_1 \\ -a_2 \end{bmatrix} \tag{4.8}$$

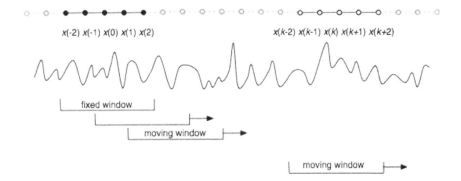

x(-2) x(-1) x(0) x(1) x(2)   x(k-2) x(k-1) x(k) x(k+1) x(k+2)

fixed window

moving window

moving window

Figure 4.1: Computation of the segmentation error measure over a sliding window. A fixed window of data is used to build a model f (in the current discussion, the model is an AR(p) model). Then this fixed model (estimated from the fixed window) is used to compute a prediction error score in each sliding window. A structural change in the dynamical behaviour of the system is identified when the score in the moving window exceeds some threshold.

Again, this is a very simple system to solve: one need only estimate $\gamma_{xx}(0)$, $\gamma_{xx}(1)$, and $\gamma_{xx}(2)$.

The general solution for AR(p) can also be obtained in the same way, as the solution of the so-called Yule–Walker equations:

$$\begin{bmatrix} \gamma_{xx}(1) \\ \gamma_{xx}(2) \\ \vdots \\ \gamma_{xx}(p) \end{bmatrix} = \begin{bmatrix} \gamma_{xx}(0) & \gamma_{xx}(1) & \cdots & \gamma_{xx}(p-1) \\ \gamma_{xx}(1) & \gamma_{xx}(0) & \cdots & \gamma_{xx}(p-2) \\ \vdots & \vdots & \ddots & \vdots \\ \gamma_{xx}(p-1) & \gamma_{xx}(p-2) & \cdots & \gamma_{xx}(0) \end{bmatrix} \begin{bmatrix} a_1 \\ a_2 \\ \vdots \\ a_p \end{bmatrix} \tag{4.9}$$

is the optimal model of the form

$$\hat{x}(t) = a_1 x_{t-1} + a_2 x_{t-2} + \ldots a_p x_{t-p} \tag{4.10}$$

(note that the structure of Eqn. (4.9) is different from that of Eqn. (4.8), and that this distinction is just due to direct algebraic manipulation).

We can now fully apply the algorithm described above. The procedure is represented graphically in Fig. 4.1. However, rather than compute the model prediction error directly, it turns out to be better to compute a model error measure based on the power spectrum. The main reason for this is that by doing so, we can determine not only whether the signal structure has changed, but if so, how it has changed. So, let $P_f(\omega)$ be the model prediction error of the model when it is applied to the fixed window data. That is, the fixed

window data is the data from which the model was built. For this data, the
model is the best model of the form (4.10). We compute the parameters of
the model, the model prediction error, and then the power spectrum of the
model prediction error. Furthermore, let $P_m^{(k)}$ be the power spectrum of the
model prediction error from applying the (*same*) model to the moving window.
Then, the segmentation error measure [28] can be computed as

$$E_k = \frac{1}{2\pi} \int_{-\pi}^{\pi} \left[P_m^{(k)} - P_f(\omega) \right]^2 d\omega. \tag{4.11}$$

Unfortunately, this is a little hard to compute. We can simplify things by
observing that the power spectrum is the Fourier transform of the autocorrelation $\gamma_{xx}(\tau)$ and this is an even discrete function. Hence

$$P(\omega) = \frac{1}{2\pi} \int_{-\infty}^{\infty} \gamma_{(xx)}(\tau) e^{-i\omega\tau} d\tau$$

$$= \frac{1}{2\pi} \left[\gamma_{xx}(0) + 2 \sum_{n=1}^{\infty} \gamma_{xx}(n) \cos n\omega \right] \tag{4.12}$$

and hence

$$E_k = \frac{1}{4\pi^2} \int_{-\pi}^{\pi} \left[\gamma_m^k(0) + 2 \sum_{n=1}^{\infty} \gamma_m^k(n) \cos n\omega - \gamma_f(0) - 2 \sum_{n=1}^{\infty} \gamma_f(n) \cos n\omega \right]^2 d\omega$$

$$= \frac{1}{4\pi^2} \int_{-\pi}^{\pi} \left[\gamma_m^k(0) + 2 \sum_{n=1}^{\infty} \gamma_m^k(n) \cos n\omega - \gamma_f(0) \right]^2 d\omega$$

$$= \frac{1}{2\pi} \left[(\gamma_m^k(0) - \gamma_f(0))^2 + 2 \sum_{n=1}^{\infty} (\gamma_m^k(n))^2 \right] \tag{4.13}$$

where γ_f and γ_m^k are the autocorrelation of the model prediction error in the
fixed and moving windows respectively.

The good thing about Eqn. (4.13) is that the expression for E_k neatly
divides into two parts. The term $(\gamma_m^k(0) - \gamma_f(0))^2$ is due to change in the
variance of the model prediction error, and the term $2 \sum_{n=1}^{\infty} (\gamma_m^k(n))^2$ determines change in the shape of the signal. Hence, by examining which of these
two terms has got larger, we can determine not only whether the signal has
changed, but also how it has changed.

Since $\gamma_f(0) = E(x(t)^2)$ over the fixed window and $\gamma_m^k(0)$ is the same quantity over the sliding window, by comparing the magnitude of these two quantities in $(\gamma_m^k(0) - \gamma_f(0))^2$ we obtain an expression for how the amplitude of
the signal and the signal noise has changed. Conversely, the second term
$2 \sum_{n=1}^{\infty} (\gamma_m^k(n))^2$ contains only the higher-order γ_m^k terms; there are no terms
describing the shape of the model prediction errors in the fixed window. That
is, $\gamma_f(n) - 0$ for all $n > 0$. The reason for this is that the model was fitted to

this data and hence, if the model is appropriate for the data (which it should be), the model prediction errors will be independent ransom numbers — the model prediction errors are uncorrelated by definition. Nonetheless, since the autocorrelation $\gamma(n)$ and the Fourier power are a Fourier transform pair (see Eqn. 4.12), then it is precisely the terms $\gamma_m^k(n)$ for $n > 0$ that determine the shape of the Fourier power spectrum of the model prediction error within the sliding window (the term $\gamma_m^k(0)$ is the DC offset). That is, the power spectrum of the model prediction error is completely described by the terms $\gamma_m^k(n)$ for $n > 0$.

4.2 Electrocardiographic Signal Processing

Now, let us consider a second example of the application of signal processing techniques to uncover features of the dynamical structure of the system. This time, we look at electrocardiographic (ECG) recordings, and the methods we will use are a little more specialised and mechanistic.

In Fig. 4.2 we see the typical sequence of phases in the standard ECG waveform. The P-wave corresponds to the electrical *depolarisation* of the upper chambers of the heart. By depolarisation we mean the phase when electrical excitation passes over the tissue, causing it to contract. Thus, the P-wave corresponds to the period of active contraction (and pumping) in the atria. The Q- R- and S-waves all correspond collectively to the orderly depolarisation of the lower chambers — that is, the pumping contraction phase of the ventricles. Hence, the P-wave indicates when blood is expelled from the atria and forced into the ventricles, and the QRS complex (as the Q-, R-, and S- waves are collectively known) corresponds to the pulse of blood being forced from the ventricles into the body's main arteries. Now, with its work done for one beat, the heart rests. During that resting phase, the heart *repolarises*. Repolarisation is the recharging of the heart tissue (much like a capacitor needs to charge) before it is able to support another wave of excitation, and another contraction. The repolarisation of the ventricles manifests as the T-wave and the repolarisation of the atria appears as the U-wave. In a normal ECG, the U-wave is usually either very difficult to detect or completely obscured by the T-wave.

Because each of these characteristic shapes corresponds to physiological activity in the body, medical practitioners have devised a series of measures of various time intervals (the more common of these are depicted in Fig. 4.2) to quantify various physiological features. For example, the RR-interval, defined as the time between successive R-waves, is a simple measure of the heart rate. The PR interval measures the time taken for the generation and conduction of electrical excitation through the atria, and the QT interval measures

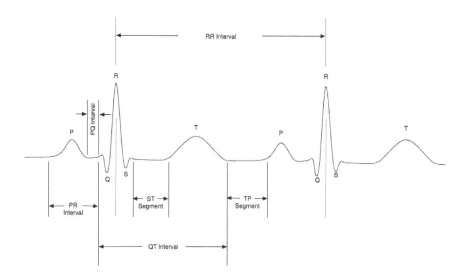

Figure 4.2: **Components of the ECG waveform.** The ECG waveform is the electrical manifestation of the regular beating of the heart. Consisting of the P-, Q-, R- S- and T-wave (along with the U-wave, often masked by T), each wave of electrical activity corresponds with a distinct physical phase of activity in the heart. From these phases, physiologists have prescribed several quantities of interest: parameters which can be measured directly from the wave form and yet relate to the underlying physical function (or dysfunction) of the heart.

propagation and repolarisation at the ventricles. In the current discussion we focus on how to construct an automated mechanism to measure just one of these quantities: the ST segment. The definition of the ST segment is a little complicated (we'll come to that later), but it is a measure of the duration between depolarisation of the ventricles and subsequent repolarisation. It measures how quickly the ventricles recover from the effort of pumping and is therefore an indication of the efficacy of blood supply to the heart muscle (that is, the blood feeding the muscles of the heart, not the blood being pumped through the heart).

Of course, the first thing we need to do if we are going to provide a numerical value for the ST segment is find the location of each heart beat. That is, to find the waves of excitation within the signal. To do this we focus on the QRS complex, as this is the most readily identifiable part of the normal ECG rhythm. There are several methods to identify the QRS complex, relying either on template matching (in one form or another) or filtering. For the template matching techniques, we revert to the concept of a sliding window introduced in Section 4.1. In Fig. 4.3 the basic idea is illustrated again. This time, it is actually a little simpler than it was in Fig. 4.1 as we now have no model to be concerned with.

The basic idea of template matching is that one somehow constructs an idealised template of the signal for which we are searching — in this case the QRS wave form. The template is then compared to short segments of the signal under study. At each time t, one computes some measure $m(t)$ of how closely the template and the segment of ECG signal match. The template continues to slide along the ECG signal and the value of $m(t)$ will continue to vary. If $m(t)$ is measuring the closeness between two signals, then the minimum is what is required (usually, this can be done by looking at the mean difference between the signal and the template). Nonetheless, it is actually more usual, and useful, to compute the correlation between the signal and the template. Hence $m(t) = \gamma_x y(t)$; and when $m(t)$ is close to unity, then the agreement between template and signal is good.

Alternatively, one can exploit the features of the QRS complex itself. The reason that the QRS complex is of interest is that it includes very large and very sudden changes in magnitude. That is, the first and second derivative of the signal should be large in magnitude. Hence, to detect this, we can do away with the need for a template and instead use some computational signal processing to derive a measure of the magnitude of the first and second derivatives. The method, described below, is that implemented in various medical-grade ECG analysis software [28]:

1. Start with the signal $x(t)$.

2. Numerically differentiate to get $y_0(t) = x(t) - x(t-1)$.

3. Numerically differentiate again: $y_1(t) = x(t) - 2x(t-1) + x(t-2)$.

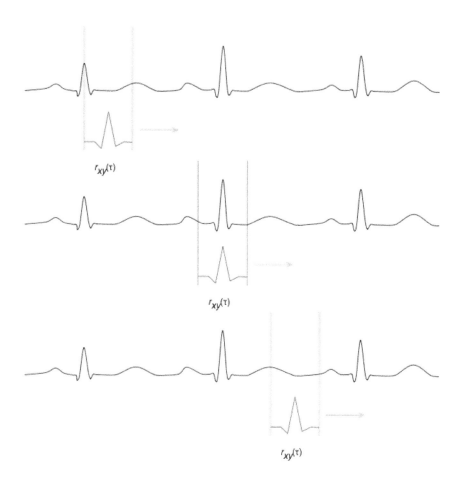

Figure 4.3: **The sliding window revisited.** The pre-defined template is progressive compared to successive short segments of the incoming ECG signal to look for good agreement. The periodic ECG signal will generate repeated periodic good agreement between the ECG signal and template.

4. Further smooth the signals $y_0(t)$ and $y_1(t)$ to obtain numerical approximations to the first and second derivatives.

5. Take a linear combination of the result $y_2(t) = 1.3y_0(t) + 1.1y_1(t)$ (of course, the exact nature of the linear combination will depend on the application).

6. Threshold the results:

$$y_3(t) = I\left[(y_2(t) > \theta) \wedge \left(\sum_{i=1}^{8} I\left[y_2(t+i) > \theta\right] > 6\right)\right].$$

7. If $y_3(t) = 1$, then t is near the R-wave peak or a QRS complex.

8. R-wave peak is located at the maximum of $x(t)$ over all t for which $y_3(t) = 1$.

Now let us look over this algorithm. Steps 2 and 3 perform the first and second differentiation. Numerically, of course, this is easy; we just subtract successive values. Of course, exactly how this should be implemented will depend on such factors as the sampling rate and whether any filtering has been done to reduce noise in the original signal (if not, such filtering could be included in the differentiation process). Nonetheless, the first derivative will be large in magnitude (first large and positive, and then large and negative) throughout the QRS complex, and the second derivative will also be large (particularly at the peak of the R-wave, when the first derivative is relatively small, but changing rapidly). Hence, by taking a linear combination of these two quantities we obtain $y_2(t)$, which should be largest from the Q- to S-wave. Finally, in step 6 we perform a thresholding operation. We seek out a point for which $y_2(t)$ is larger than some threshold θ and remains so for six of the next eight time steps. Note that the function $I(X)$ is defined by

$$I(X) = \begin{cases} 1 & \text{if } X \text{ is true} \\ 0 & \text{otherwise} \end{cases},$$

and the symbol $\wedge$ denotes the usual logical operation of AND.

Now, by either applying this QRS filtering method or one of the template matching methods, we can identify the components of the wave form. In particular we can expect to correctly locate the peak of the R-wave. We therefore have a section of time series data like that depicted in Fig. 4.4. More than likely this will actually be obtained as an ensemble average of several sections of data, to reduce the effects of noise. Nonetheless, we can now proceed to describe an algorithm — which can be automated — to detect the ST segment [28]. Conveniently, we will claim that this algorithm will also serve as a *definition* of the ST segment. But, we loosely mean that the ST segment is the time taken for the ventricles to repolarise.

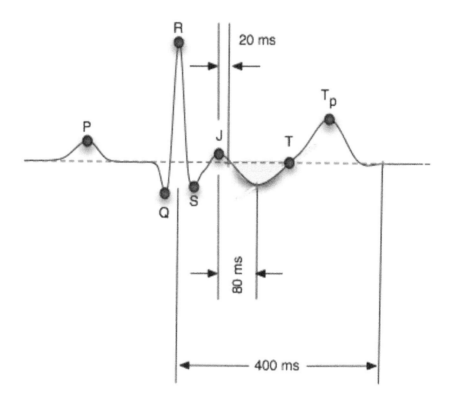

Figure 4.4: Identifying the ST segment. A description of the algorithm is given in the text. In addition to identifying the ST-interval, one can also extract other physiologically useful parameters such as the ST area (shaded region) and the ST slope (drawn as a line segment from the shaded region).

1. Let P denote the peak of the P-wave, Q the trough of the Q-wave, R the peak of the R-wave, and S the trough of the S-wave. Hence, P, Q, R, and S may all be regarded as Cartesian co-ordinates (and similarly for the points that follow).

2. Let J denote the first point of inflection at or after S. That is, J is the first point following S for which the first derivative is zero. In cases where S is a point of inflection and not a true minimum, J will be at S.

3. Find the flattest (closest to zero slope) line segment of 30 ms between P and Q. This line is defined to be 0 voltage.

4. Let T_p denote the peak of the T-wave such that T_p is between $J + 80$ ms and $R + 400$ ms. Let T be the point prior to T_p closest to the zero voltage defined in step 2.

5. The duration from J to T is defined as the ST-interval.

Thus, knowledge of the dynamics of the underlying system, exploitation of electrical properties of the system for measurements and computational analysis of the resulting signal allows us to provide medically useful information about the state of the biological system — both in Sections 4.2 and 4.1. In the remainder of this chapter we turn our attention to how simple measurement, together with basic mathematical principles, can be used to acquire information about the underlying system.

4.3 Vector Cardiography

In Sec. 3.2 we learnt how multiple ECG leads are used to obtain distinct scalar measurements of the underlying electrical potential (a vector). The heart, of course, is a three-dimensional object; and when we measure electrical potential of the heart, we are really measuring the average electrical potential in a given direction — corresponding to the specific ECG lead which we choose to look at (Section 3.2) — over the heart. Hence, the need to employ multiple different ECG leads to obtain a complete picture of electrical activity over the entire three-dimensional mass of the heart. The signal processing methods described in Section 4.2 deal only with a scalar time series and how to extract quantitative measurements of the specific features observed in that signal. It turns out that cardiologists also have standard techniques with which to reconstruct the underlying ECG *vector* time series.

Essentially the method is straightforward and amounts to nothing more than a graphical re-representation of the information already present in a multi-lead ECG recording. There are many different ways of obtaining a

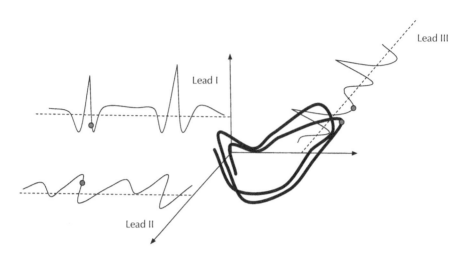

Figure 4.5: Vector ECG reconstruction. The three scalar time series (in this example, Lead I, II and III) are shown as light lines. Combining these we obtain a vector time series, shown as a heavy line. At any instant in time, the current state measured with each of the scalar time series is a single value; these combine to give the three-dimensional points (all shown as small circles).

vector cardiogram, but all are essentially the same. Suppose that $v_1(t)$, $v_2(t)$ and $v_3(t)$ are three separate ECG leads. These may be leads I, II and III; some other combination of the standard leads or some non-standard lead combination. In any case, by simply combining them, it is possible to obtain a vector time series

$$v(t) = [v_1(t), v_2(t), v_3(t)], \tag{4.14}$$

and hence the ECG signal traces out a curve in three-dimensional space, as depicted in Fig. 4.5. The three-dimensional shape in Fig. 4.5 represents the overall behaviour of the system. For a given state (for example, shown as a dot in this figure), one can trace the future evolution of the system behaviour by following the trajectories. This is exactly equivalent to the phase plane concept of Chapter 2: By taking multiple scalar time series and plotting the values together as a vector quantity, one obtains a high-dimensional phase plane, the so-called *state space*.

For the practising cardiologist, this three-dimensional representation of the ECG represents an alternative way of visualising the data, and potentially diagnosing an ailment. From the viewpoint of the dynamical behaviour of the system, this picture offers a succinct description of the future behaviour of the system and characterises its deterministic structure. Yet, this method of creating a geometrical high-dimensional object from multiple (or even a single) scalar time series is not restricted to the ECG.

4.4   Embedology and State Space Representation

We can extend the basic principle of vector reconstruction from scalar measures of Section 4.3 to arbitrary time series data. Doing so will give us power techniques which can be applied to a wide variety of data [34]. Let $x(t)$ represent the state of the system at time t. The process of taking a scalar time series $x(t)$ and, through some transform, obtaining a vector time series $v(t)$, which is mathematically equivalent to the underlying state of the system in a well-defined state space, is called *embedding*.

Now, suppose that x_t is a variable measured from a d-dimensional dynamical system. Such deterministic systems can be described by a set of d differential equations: either d first-order equations in d different variables, or d-order differential equations in one variable. In other words, to completely describe the current state of the system, we need x_t and its first $(d-1)$ derivatives:

$$x(t) \rightarrow \left[x(t), \frac{dx}{dt}(t), \frac{d^2x}{dt^2}(t), \dots, \frac{d^{d_e-1}x}{dt^d}(t) \right]. \qquad (4.15)$$

In the most basic case, one only has access to a single scalar measurement — varying in time. If we wish to reconstruct the system state space from that measurement, we need to be able to estimate the various derivatives. Of course, each of the derivatives in Eqn. (4.15) can be estimated numerically,

$$\frac{dx}{dt} = \lim_{\delta \to 0} \frac{x(t+\delta) - x(t)}{\delta},$$

and so on. Hence, the embedding described by Eqn. (4.15) actually includes information of $x(t)$ at successive (or equivalently, preceding) times: $x(t), x(t+\delta), x(t+2\delta), \dots$. Naturally, this means that is it possible to reconstruct the system state space with a *time delay* embedding:

$$x(t) \rightarrow [x(t), x(t-\delta), x(t-2\delta), \dots, x(t-(d_e-1)\delta)]. \qquad (4.16)$$

Now, the problem is two-fold. We must know what are appropriate values for both the *embedding dimension* d_e and the *embedding lag* τ.

Although there are rigorous mathematical foundations which guarantee that the transformation (4.16) is a true embedding, and can be used to represent the underlying state of the system[2], there is no such rigorous pronouncement

---

[2] Under certain fairly generic conditions.

regarding the proper estimation of either d_e or τ. Rather, mathematical theory says that $\tau = 1$ (whatever the sampling rate) is sufficient and that d_e is correct when d_e is large enough[3].

Actually, the situation is not that bleak. Although differing values of both d_e and τ will give different results that will certainly be affected by the data, the dependence on these values is not critical. Usually, there will be a wide range of values of τ which are sufficient and the criteria for suitable values of d_e are (in theory at least) only a lower bound — any value above that minimum value will work equally well (provided there is sufficient data). Of course, the choice of values will also depend on the purpose for which the embedding is designed. A value which is suitable for estimating correlation dimension (Section 4.5) is different from what is required for building a predictive model (Section 4.6).

First, let us consider a suitable choice for the embedding dimension d_e. Suppose that the purpose of the transform (4.16) is really construction of a proper embedding. The basic requirement of such an embedding is that trajectories will not cross. As we have seen in the previous chapters, the current state of the system is sufficient to describe its future behaviour. However, if two trajectories were to cross, then that future would not be unambiguous. Hence, we simply choose d_e such that no trajectory crosses another either orthogonally or at an oblique angle (tangential almost-crossings can be permitted as this would indicate two trajectories that were, nonetheless, still going in the same direction). Figure 4.6 explains.

One of the most straightforward and widely accepted ways of calculating whether, for a given value of d_e, such non-tangential intersections are occurring is the so-called method of false nearest neighbours [34]. Neighbours of the current state in state space are those other states which are close. Now if these neighbours really do correspond to true neighbours in the original mathematical state space, then they should stay close as we increase the embedding dimension. The method of false nearest neighbours merely calculates the number of points that appear to be neighbours in dimension d_e, but are not neighbours in dimension $d_e + 1$. This number will decrease as d_e increases. When it becomes negligible, then we have found a suitable (sufficient) value of d_e.

Second, choice of embedding lag is both relatively more straightforward but also potentially more complex. Choice of embedding lag is straightforward because for many purposes, almost any value will do (as predicted by mathematical theory). However, when faced with the practical problems of limited experimental data contaminated with measurement noise, the issue of choosing which value is best becomes more complex. Nonetheless, there are

---

[3]More precisely, if d_e is too small, the transform (4.16) will not give a proper embedding and this should be evident. One should increase d_e until an embedding, of something, results. Unfortunately, one cannot address the possibility of a non-unique and incorrect embedding transform which may occur for a finite experimental data set.

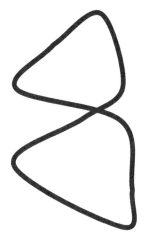 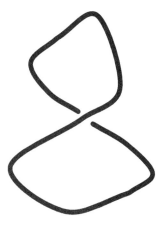

Embedding dimension: 2   Embedding dimension: 3

Figure 4.6: **Embedding dimension**. The embedding dimension should be high enough so as to avoid self-intersection of trajectories. At any instant in time, the state of the system is defined by a single point on this curve. As time passes, that state moves along the curve. However, in the two-dimensional version on the left, there is ambiguity about the future behaviour of the system at the central crossing point; the figure-eight trajectory cannot be embedded in two dimensions without an ambiguous self-intersection. Three dimensions are required (as we have attempted to depict on the right).

several competing criteria to help make that decision. The principle of each of these techniques is to choose a value of τ which will allow for the maximum amount of information about the underlying system to be expressed in the embedding (4.16).

Hence, the simplest of these criteria is to choose τ equal to the first zero, or if not, then the first minimum of the autocorrelation function,

$$\begin{aligned}\gamma_{xx}(\tau) &= E(x(t)x(t+\tau)) \\ &= E(x(t)x(t-\tau)),\end{aligned} \tag{4.17}$$

that is, Eqn. (4.5). Alternatively, some researchers have suggested choosing a decorrelation time — that is, the time when the autocorrelation falls below a specific level (most usually $\frac{\gamma_{xx}(0)}{e}$). However, a strong objection to using the autocorrelation function is that, although it is simple, it is also strictly linear and can therefore not properly measure nonlinear correlation. Hence, an alternative is to use the minimum of the *mutual information* function which is defined by

$$I(X,\tau) = \sum_{x(t)}\sum_{x(t+\tau)} p(x(t),x(t+\tau))\log\left(\frac{p(x(t),x(t+\tau))}{p(x(t))p(x(t+\tau))}\right), \tag{4.18}$$

where $p(x(t))$ is the probability of observing $x(t)$, and $p(x(t),x(t+\tau))$ is the joint probability of observing both $x(t)$ and $x(t+\tau)$. While mutual information (4.18) is more general and probably theoretically "better", it can also be rather difficult to estimate as one must estimate both the probabiility distribution of the observed variable $x(t)$ and also the joint probability distribution $p(x(t),x(t+\tau))$. This, in turn, can depend on the choice of various estimation parameters. In light of this, one final, expedient, alternative is to take

$$\tau = \frac{1}{4}(\text{period}). \tag{4.19}$$

If the time series $x(t)$ exhibits some periodic fluctuation, then the advantage of this method is that the period is usually fairly easy to estimate. If the signal itself is exactly linear, then this is equivalent to taking the first zero of the autocorrelation function. However, for nonlinear signals, this approach can also take account of more subtle time varying behaviour without the need for estimation of joint probability distribution functions as in (4.18). Figure 4.7 gives an example of the calculation of these embedding parameters.

Hence, using some combination of the methods described above, we have obtained suitable values of τ and d_e. Using these numbers we are able to construct a mapping (4.16) which forms a time-delay embedding of the original system. The vector state variable $v(t)$, given by

$$v(t) = [x(t), x(t-\delta), x(t-2\delta)\dots, x(t-(d_e-1)\delta)], \tag{4.20}$$

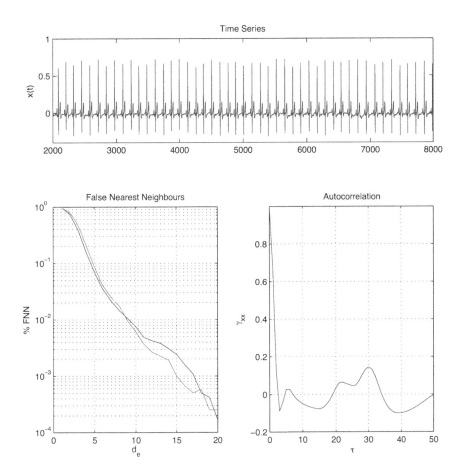

Figure 4.7: **Calculating embedding parameters**. From the ECG time series shown on the top, we calculate the proportion of false nearest neighbours (with different embedding lag $\tau = 1, 2, 3$) and the autocorrelation (which indicates an optimal lag $\tau = 3$).

Figure 4.8: **The embedded state space**. Representation of the embedded state space for the ECG time series data from Fig. 4.7 ($d_e = 3$ and $\tau = 3$). The large solid dots represent the discrete states of the time series, and the faint dotted line connects successive states. See colour insert.

is now equivalent to the original state of the underlying deterministic dynamical system: described by some unknown system of d_e (or fewer) differential equations. Hence, from the set of points $\{v(t)|t = 1, 2, 3, \ldots\}$ and the corresponding deterministic mapping $v_t \rightarrow v_{t+1}$, we can learn everything there is to know about this system. Figure 4.8 depicts the state-space embedding of the data from Fig. 4.7. In Section 4.5 we briefly review some quantitative measures which can usefully be estimated from the set of points. In Section 4.6 we introduce one approach to estimating the mapping $v_t \rightarrow v_{t+1}$.

4.5   Fractals, Chaos and Non-Linear Dynamics

The methods discussed in Sections 4.1 and 4.2 relied heavily on the tools of linear signal processing. That is, the information that these methods can extract from the underlying signal is inherently restricted to linear qualities. This is like taking the difference and differential equations discussed in Section 1.2 and Chapter 2 (and later in this text) and only considering the linearisation of those equations. In certain situations (for example, near a fixed point with dominant linear terms), this can be useful. However, this is not always the case. Moreover, in some very important examples (for example, the chaotic dynamics of the logistic map), the *only* important thing is system non-linearity.

Hence, when we look at time series data in an effort to characterise the system, there is also a need to be able to quantify the nonlinear properties of the system. One of the simplest of such properties, and the one which we will expend most effort discussing here, is the *correlation dimension*. For a more extensive discussion of other related instruments, see, for example, [34]. If a system is exhibiting a strictly periodic dynamical behaviour, then the system behaviour repeats over and over again. In state space, the path inscribed by a periodic system will always be a closed loop — like the figure-eight of Fig. 4.6. The closed loop, of course, is a one-dimensional object. Similarly, a superposition of two (incommensurate) frequencies will result in a dense torus — a two-dimensional object. And so, every integer dimension corresponds to a fixed integer number of incommensurate frequencies[4].

In contrast, if a system were to be completely random, then the system state would move about the state space completely randomly. If we were to trace such a trajectory, it would appear as a dense scribble gradually filling up all available space; in such a situation, there would be no embedding dimension capable of avoiding all false nearest neighbours (remember that false

---

[4]The frequencies need to be incommensurate. This means that the frequencies f_1 and f_2 cannot be expressed as any rational multiple of each other.

Figure 4.9: Fractal broccoli. Romanesque broccoli is one of the more striking examples of a naturally occurring fractal. The self-similar repeating geometric patterns are the hallmarks of a fractal. Consequently, the roughness of the surface of the vegetable is somewhere between two and three dimensional. (Image reproduced from http://en.wikipedia.org/wiki/File:Brassica_romanesco.jpg)

nearest neighbours arise from the assumption that the underlying dynamics are *deterministic*). In this case, the dimension of the dense scribble would be close to the embedding dimension — it fills the available space.

What about chaotic dynamical systems? By definition (see Section 2.3), such a system is neither random nor periodic. Yet, since it is also bounded, the trajectory must remain constrained within some finite state space. In such situations, the dimension of the space occupied by trajectories is actually fractional[5]. The existence of such fractional dimensions can be tested by estimating the correlation dimension. Objects exhibiting a fractional correlation dimension are known as *fractals*. While it is not important for the present discussion, it is worth mentioning that such things occur surprisingly

---

[5]Mostly, but not always — see [34].

frequently in natural systems — structures which exhibit complex patterns repeated across many spatial scales, see Fig. 4.9.

However, rather than looking at the surface of a head of broccoli, we are interested in the distribution of the set of states in state space — for example, the points depicted in Fig. 4.8. Let x_i and x_j represent randomly chosen points, from among a set of N states, sampled from the state space reconstructed from a time delay embedded of a single trajectory of some deterministic system[6]. Then the correlation integral is defined to be

$$C_I^{(N)}(\epsilon) = \text{Prob}(\|x_i - x_j\| < \epsilon) \tag{4.21}$$
$$= \frac{1}{N(N-1)} \sum_{i,j,i\neq j} I(\|x_i - x_j\| < \epsilon),$$

where $I(X)$ is the indicator function and has a value of 1 if X is true and zero otherwise. The correlation dimension is defined to be

$$d_c = \lim_{N\to\infty} \lim_{\epsilon\to 0} \frac{\log C_I^{(N)}(\epsilon)}{\log \epsilon}, \tag{4.22}$$

which is the slope of the correlation integral in the limit of very small lengths, ϵ. Figure 4.10 gives an example of this calculation for the ECG data from the previous section (Fig. 4.7 and Fig. 4.8) — in this case yielding a correlation dimension of about 1.3.

The actual numerical value one obtains for correlation dimension is not, of itself, particularly important — it depends to some extent on the actual choice of some parameters. Nonetheless, its relative value can be important. By comparing correlation dimension values for different data sets, one can distinguish (for example) between normal and arrhythmic heart activity [41, 50] or identify coronary risk patients from among a group of normal subjects by observing only their normal ECG rhythm [52, 53]. There are many other examples, with both correlation dimension as well as other non-linear statistics (such as Lyapunov exponents, computational complexity and entropy [34]) applied to many different biological[7] systems.

4.6 Prediction

In the previous section (and in Section 4.4), non-linearity in the underlying system is exploited to provide new statistical tools — new things we can measure from data, which, hopefully, are useful for diagnostic or other practical

---

[6]There is no real reason why these points must come from a single trajectory.

[7]And also non-biological.

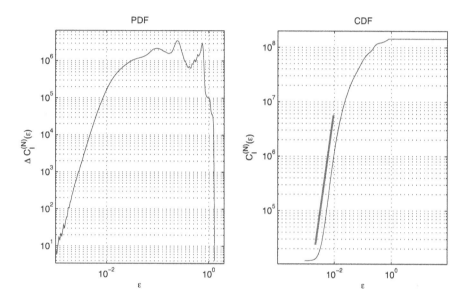

Figure 4.10: Calculation of correlation dimension. Calculation of correlation dimension for the data in Fig. 4.7 (with $d_e = 5$ and $\tau = 3$ as suggested by Fig. 4.7). On the left, the probability distribution function for the correlation integral (that is, the probability that two points are a specific distance apart, in this case, with logarithmic distribution of bins); and on the right, the correlation integral (the probability that two points are less than a specific distance apart). In both panels, the vertical axes are actual counts rather than properly normalised probabilities, and therefore not important. From the right-hand (CDF) plot we can compute the correlation dimension as the slope of this curve as ϵ gets small. However, the extreme tail of the distribution is not relevant as this represents truncation due to quantisation of the recorded data. Hence we fit the straight line (shown as a heavier line) on the apparently straight region which is larger in value.

concerns. But, there is another important reason for performing the embedding procedures outlined in Section 4.4. In many cases it is not possible to write down the differential or difference equations that govern the behaviour of a particular system — often we just do not know enough about the biology to provide an adequate mathematical description. Nonetheless, even in these situations, we want to be able to describe the dynamics.

The primary purpose for a differential or difference equation description is to predict the future values of that system. Once we can predict the future, it is then possible to locate and study the fixed points, stability and bifurcation behaviour of the system. Hence, our objective here is to find a function F such that $F(v_t) \approx V_{t+1}$, where v_t is the vector values state of the system at time t. If we denote $d_w = d_e \tau$ as the embedding window, and note that at time t we have knowledge of every preceding state $x_t, x_{t-1}, x_{t-2}, x_{t-3}, \ldots$, then most of v_{t+1} we already know. The new state we wish to predict, v_{t+1}, can already be written as

$$v_{t+1} = [?, x_{t+1-\tau}, , x_{t+1-2\tau}, \ldots, , x_{t+1-(d_e-1)\tau}],$$

where only the first component of the right-hand side is unknown. Hence the problem is actually to find a function f such that

$$f(x_t, x_{t-1}, x_{t-2}, \ldots, x_{t-d_w+1}) = x_{t+1} + \epsilon_{t+1}, \tag{4.23}$$

where the expected prediction error $E(\epsilon^2)$ is minimised. If we strictly follow the embedding procedure of Eqn. (4.20), only some of the components of the function f (namely $x_t, x_{t-\tau}, x_{t-2\tau}, \ldots, x_{t-(d_e-1)\tau}$) are actually used in f.

There are many solutions to construct a function of this form. The one approach that we consider here is to use radial basis functions, but neural networks, support vector machines, or one of many alternative approaches to computational function approximation will work (potentially) equally well. Nonetheless, whatever functional approach one chooses to use, there are some basic similarities. Using some collection of function $\phi(\cdot)$, we attempt to construct a function which achieves the desired result — minimising $E(\epsilon^2)$ in the simplest way possible.

Finding f is now a problem of finding the best function,

$$f(v) = \sum_{i=1}^{m} \lambda_i \phi \left(\frac{\|v - c_i\|}{r_i} \right), \tag{4.24}$$

which is a linear combination (the λ's) of nonlinear functions ϕ, where each function has a set of parameters c_i and r_i describing its specific behaviour. Standard algorithms exist to do all this: The parameters λ_i ($m = 1, 2, \ldots m$) can be solved by using linear optimisation (essentially, this is just the least squares technique reported in Section 4.1 for linear EEG signal analysis); and the parameters c_i and r_i can be calculated with either a non-linear optimisation technique or a guess-work heuristic. Essentially, by "guess-work heuristic" we mean applying some clever guesswork to generate a large number of possible candidates (c_i, r_i) and then choose whichever one works best. For radial basis models, such methods have been described extensively in many places, for example [15, 35].

When Eqn. (4.24) is a radial basis function, the individual basis functions $\phi(\cdot)$ have the specific form of a smooth (that is, differentiable) function which decreases to zero[8]. One popular and simple example is the Gaussian basis function

$$\phi(x) = \exp \frac{-x^2}{2}.$$

In this case, the two parameters r_i and c_i have special meaning. The parameter c_i is the *centre* and defines the location of that particular basis function. The second parameter r_i is the *radius* and defines the size of the function.

However, there remains one parameter in Eqn. (4.24) which is rather more problematic: m. The aim is to estimate the parameters of Eqn. (4.24) from some observed data in such a way that the error is minimised. However, increasing m will increase the total number of parameters and therefore make it easy to reduce $E(\epsilon^2)$ arbitrarily. Imagine increasing m so that there are more parameters than data (a rather extreme example). This would mean that the function f could fit the observed data perfectly, but, at the same time it would be useless to try to generalise from f the behaviour of the system on any new data. Conversely, if m is too small, f will not be able to capture all the important features of the observed data. Some balance between the two extremes needs to be found.

One rather common way to achieve this is to split the observed data into two sections: the *fitting* data and *testing* data. Using only the fitting data, one builds a model for a fixed value of m. The procedure is repeated with different values of m and the same fitting data. Then each model is applied to the testing data, and the model with the best performance on the testing data is chosen to be the best model (or best m). This method is practical if there is abundant data. However, for situations where the available data is limited, information theoretic criteria need to be employed instead. Effectively, these methods estimate (with various statistical techniques) the compression one achieves with the model. The model which gives the most compact description of the data is judged to be the best.

For example, if no model is used, the cost of describing the data is just the cost of listing all the data values. However, if some model is used, one can describe that model, plus the model prediction errors (which one assumes will be smaller and easier to describe than the original data). If the model is good and not too big, then this option will give the shortest description of the data. This is the approach we adopt to build the modesl shown in Figs. 4.11, 4.12, and 4.13. As these figures clearly indicate, the model behaves well in

---

[8]Strictly speaking, it should have *compact support*. This means that the integral $\int_0^\infty \phi(x)dx$ should be finite.

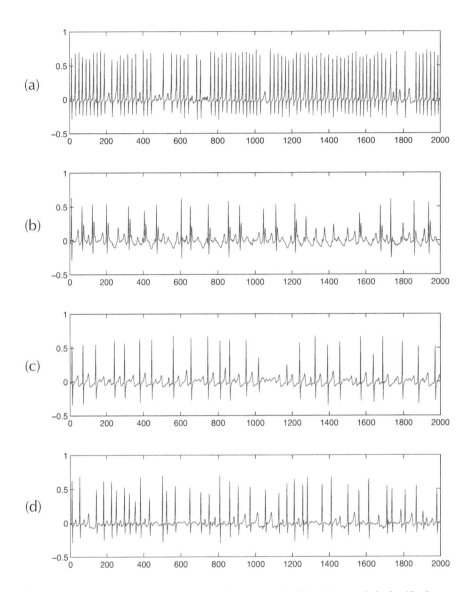

Figure 4.11: **Behaviour of non-linear radial basis models built from experimental ECG recordings**. Dynamical behaviour of four models built from the data in Fig. 4.7 are shown. While the first model (a) does not capture the same dynamics, it exhibits behaviour something like ventricular tachycardia — the P and T waves are completely absent. The third (c) and fourth (d) models in particular are more successful and capture the same range of dynamics as the original system.

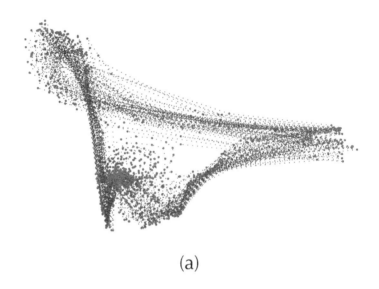

(a)

Figure 4.12:   **Embedding of a trajectory from non-linear radial basis model (a) built from experimental ECG recordings.** In comparison to Fig. 4.8, it is clear that this model lacks the complex detailed motion corresponding to the P- and T- waves. Nonetheless, it captures the QRS complex dynamics well. See colour insert.

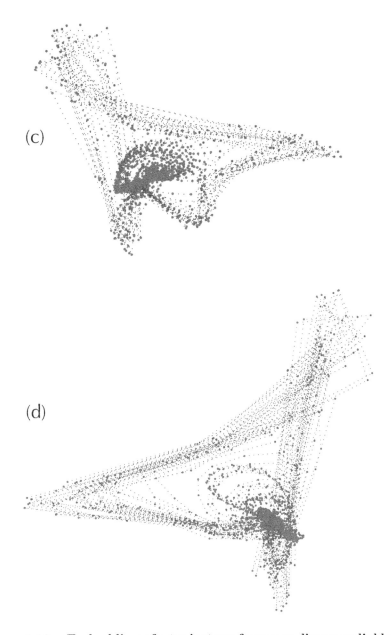

Figure 4.13: **Embedding of a trajectory from non-linear radial basis model (c) and (d) built from experimental ECG recordings — model (b) is similar but not shown here.** In comparison to Fig. 4.8, these models reproduce the same dynamical structure well — the relative coarseness of these model trajectories in comparison to Fig. 4.8 is only due to the much shorter data segment used in this simulation. See colour insert.

producing dynamics that mimic closely the dynamics of the original system (in Fig. 4.7). Now we can use this model to study the stability, fixed points and behaviour of the underlying dynamical system, even though our biological understanding of the cardiac system is currently insufficient to write down the corresponding differential equations correctly.

Finally, we note that we could now apply this type of methodology to the problem of segmentation of an arbitrary signal — and the extension of the EEG segmentation method of Section 4.1. By building a model such as in Eqn. (4.24), we have obtained a characterisation of the underlying dynamics of the system from which the data comes. We can compute the model performance $E(\epsilon^2)$ on new data as it comes in and thereby evaluate when the model behaviour changes.

4.7   Summary

The focus of this chapter has been on data processing methods. In Chapter 3 we discussed ways in which biomedical signals can be collected. In this chapter we discussed a wide variety of methods to process such data. The ECG and EEG signal processing approaches described in Sections 4.1 and 4.2 are widely accepted and used in practical patient monitoring systems. Of course, such systems need to be practical, and practical systems need to be as simple as possible: hence, these approaches are essentially linear.

Unfortunately, by restricting the data analysis to linear methods, something is always lost. In Sections 4.5 and 4.6 we described some fully non-linear approaches to understand observed time series data. While the linear methods are inherently statistical (and the main mathematical tool we employed was linear correlation), the non-linear methods assume an underlying determinism. That determinism is uncovered by using the time delay embedding (Section 4.4), which we showed can be thought of as nothing more than a generalisation of the idea of vector cardiography (Section 4.3). While new research is showing areas where these methods may be useful[9], it remains to be seen whether these new methods will be as indispensable as the standard linear techniques of EEG segmentation and ECG signal analysis.

---

[9]For example, my own most recent research has applied these modelling techniques to experimental recordings of synaptic transmission and thereby shown that synapses exhibit complex non-linear frequency dependent responses to input stimulus [37]).

Glossary

Correlation dimension A mathematical description of the distribution of points in space. For a two-dimensional object, the number of points in a square of length ℓ is ℓ^2. For a three-dimensional object, the number of points in a cube of length ℓ is ℓ^3. For a fractal, the number of points in a volume with linear length ℓ is proportional to ℓ^{d_c}, where d_c is the correlation dimension. Also known as the *fractal dimension.*

Correlation integral A mathematical description of the distribution of a set of points — it is the probability that two points are closer than a distance ϵ, expressed as a function of ϵ.

Differential embedding An embedding dependent on explicitly estimating the higher-order derivatives.

Embedding The process of taking a scalar time series $x(t)$, and through some transform obtaining a vector time series $v(t)$ which is mathematically equivalent to the underlying state of the system.

Embedding dimension The Euclidean dimension d_e of a time delay embedding.

Embedding lag The time lag τ between successive components of a time delay embedding.

State space The high-dimensional analogue of a phase plane. Each point in the state space describes a particular state of the system. Dynamical transition between states is described by movement between positions in the state space.

Time delay embedding An embedding obtained by constructing a vector state from a scalar measurement of the system at a given time and also at preceding times.

Exercises

1. How much (uncompressed) data is produced during an 8 hour EEG and ECG sleep study? State assumptions.

2. If the amplifier gain of an incoming EEG signal is changed by a factor of 2 (i.e. signal amplitude doubles), but the original signal is otherwise unchanged, what would be the value of the segmentation error measure?

3. Show, from the definition of covariance, that the signal autocorrelation is given by

$$\gamma(\tau) = E[x(t)x(t+\tau)].$$

Hence, deduce that $\gamma(\tau)$ is an even function and that the global maximum occurs at $\tau = 0$.

4. Show that, for a second-order linear predictor $\hat{y}(n) = a_1 y(n-1) + a_2 y(n-2)$ that the optimal model parameters satisfy

$$\begin{bmatrix} \xi \\ 0 \\ 0 \end{bmatrix} = \begin{bmatrix} \gamma(0) & \gamma(1) & \gamma(2) \\ \gamma(1) & \gamma(0) & \gamma(1) \\ \gamma(2) & \gamma(1) & \gamma(0) \end{bmatrix} \begin{bmatrix} 1 \\ -a_1 \\ -a_2 \end{bmatrix}$$

where ξ is the sample mean error and $\gamma(\tau)$ is the sample autocorrelation with lag τ.

5. Explain why the first term in the segmentation error measure

$$E_k = \frac{1}{2\pi} \left[(\gamma_m^k(0) - \gamma_f(0))^2 + 2 \sum_{n=1}^{\infty} \gamma_m^k(n)^2 \right]$$

relates only to magnitude of error, and the second term is determined by the waveform shape. Moreover, describe why the infinite sum $\sum_{n=1}^{\infty} \gamma_m^k(n)^2$ is expected to converge.

6. Implement the segmentation algorithm described in lectures. How does change of threshold θ affect your results? I.e. Use the MATLAB® program eeg_segment.m to segment a sample of EEG data. If you do not have access to raw data EEG data, then download the sample data eeg1.dat or eeg2.dat from the website. In either case, describe your results and justify the parameter values required to achieve your results. Determine how sensitive the results are to changes in those parameter values. In giving your answer, be sure to adequately describe the output of the simulation.

7. The segmentation error measure for a certain stationary EEG waveform has a residual noise power of 1.

   (a) If the amplification is altered so as to triple the signal amplitude (but everything else remains unchanged), what would the segmentation error measure be?

   (b) If the signal changes from α-dominated to β-dominated rhythms, describe (qualitatively) what you would observe in the segmentation error measure.

   (c) If the length of the sliding window is changed (but the signal remains as described initially), what effect would this have on the segmentation error measure?

8. A fourth-order autoregressive model is constructed to perform segmentation analysis on an observed EEG signal. The values of autocorrelation computed on the reference signal error and five moving windows are given in the following table. If segmentation is to be applied with a threshold of 1, at which window will segmentation occur?

τ	0	1	2	3	4	5	6	7	8
$r_f(\tau)$	1	0	0	0	0	0	0	0	0
$r_m^{(1)}(\tau)$	1	0.3	0.2	−0.1	−0.2	0.3	0.1	0	0
$r_m^{(2)}(\tau)$	1.2	0.4	0.3	0	−0.4	0.5	0.2	0.1	0
$r_m^{(3)}(\tau)$	1.5	0.6	0.4	0.2	−0.4	−0.8	0.5	0.2	−0.2
$r_m^{(4)}(\tau)$	1.8	0.8	0.5	0.3	−0.6	−1.1	0.3	0.5	−0.3
$r_m^{(5)}(\tau)$	2.4	1.2	0.7	0.1	−1.1	−1.5	−0.5	−0.3	−0.8

9. Suggest suitable values for the segmentation algorithm eeg_segment to successfully segment the data eeg1.dat. Both data and segmentation schemes are available from the website for this book.

10. Using the EEG segmentation algorithm described in lectures (a MATLAB® implementation is provided on the website), demonstrate segmentation of the signal eeg1.dat (also available from the website).

11. Suppose that for a particular EEG signal, the best linear predictive model is determined to be of the form $\hat{y}(n) = \alpha y(n - 1) + \beta y(n - 10)$. Derive analytic expressions for the optimal values of α, β and $\xi = E((y(n) - \hat{y}(n))^2)$.

12. Compute the segmentation error measure for the data eeg_rm1.dat, eeg_rm2.dat, eeg_rm3.dat and eeg_rm4.dat given the reference data eeg_rf.dat (all available from the website). You should use a window size of 100 and build an autoregressive model of order 6. Explain your computation.

13. Several algorithms have been introduced for the purpose of QRS complex identification. Why is template subtraction more prone to noise than template matching with cross-correlation?

14. Compare and contrast the filter-based and width detection methods for QRS detection. Suppose one is unable to identify the shape of the QRS template (or that the shape matches poorly with the observed data). In this situation, how would template matching methods compare to the differentiation techniques. How might these problems be overcome?

15. Suppose ECG is recorded using Lead II. However, inadvertently, the electrode inputs to the differential amplifier are reversed. What effect will this have on the observed signal? Moreover, how will this affect QRS detection with each of the four methods described above?

16. Show that template subtraction is more prone to noise and non-stationarity than cross-correlation when applied to QRS detections.

17. Comment on the degree to which each of the four QRS detection schemes are apparently parameter dependent. What effect is this likely to have on actual realisations of these systems?

18. Derive pseudo-code, and hence implement the ST detection scheme.

19. Implement the QT detection scheme described in lectures and apply it to the sample data `ecg.dat`.

20. Download the signal `eeg1.dat`, `eeg2.dat` and `eeg3.dat` from the book's website.

    (a) Using MATLAB® , or otherwise, compute and plot the autocorrelation (for time lag $\tau = 0, 1, 2, \ldots, 20$) for each of these three signals.

    (b) Using your answer in part (a), compute the best linear predictor of order $p = 3$ for the sample `eeg1.dat`

    $$\hat{y}(n) = a_1 y(n-1) + a_2 y(n-2) + a_3 y(n-3)$$

    (you should clear state the values you obtain for a_1, a_2, a_3 and ξ).

    (c) Compute the autocorrelation of the model prediction errors $\hat{y}(n) - y(n)$ for `eeg1.dat`.

    (d) Apply the model obtain in part (b) to the signals `eeg2.dat` and `eeg3.dat` and compute (in each case) the autocorrelation of the signal prediction error.

    (e) Finally, compute the spectral error measure for `eeg2.dat` and `eeg3.dat` using the model from `eeg1.dat` (state the order of approximation which you use for the second term).

21. The best linear fourth-order autoregressive model for an incoming EEG model is given by the general form

    $$\hat{y}(n) = a_1 y(n-1) + a_2 y(n-2) + a_3 y(n-3) + a_4 y(n-4).$$

    (a) For a given EEG signal, suppose that the optimal values for a_1, a_2 and a_3 are already known. Provide a direct derivation of the optimal value for a_4.

    (b) Suppose that $a_1 = 0.8$, $a_2 = 0$ and $a_3 = -0.3$. Provided the following table of autocorrelation values, compute a numerical value for a_4.

n	0	1	2	3	4	5
$\gamma(n)$	12.2	8.3	1.9	-3.5	-1.6	0.1

22. The spectral error measure is given by

$$E_k = \frac{1}{2\pi}\left[(r_m^k(0) - r_f(0))^2 + 2\sum_{n=1}^{N}(r_m^k(n))^2\right],$$

where $N \approx 4$, r_m^k is the autocorrelation of the model prediction error on a moving window, and r_f is the autocorrelation of the model prediction error of a fixed window on which the model is built.

(a) Suppose an initial incoming EEG signal is predominantly δ-wave and the optimal linear predictor is built from that data. If the data is stationary with mean square model prediction error $\sigma^2 = 1.3$, what is the segmentation error observed in a subsequent section of deep (stage 3-4) sleep? State any assumption you need to make.

(b) Subsequently the sleep state changes to a lighter sleep state (θ-wave). If the autocorrelation of the model prediction error in this new sample is given in the table below, compute the new value of spectral error.

n	0	1	2	3	4	5
$\gamma(n)$	16.2	6.3	-1.9	3.8	-1.2	-0.1

(c) How would you interpret the difference between the answer in the previous part questions? Which term in the spectral error measure dominates? Does this reflect mostly change in amplitude or shape?

Chapter 5

Computational Neurophysiology

5.1 The Cell

In the next two chapters, we take a much closer look at mathematical modelling of a particular biological system. The system we choose to examine is perhaps the most basic of all such systems – a single cell. In this chapter we will focus on the necessary physiological background; in the next we turn to the mathematical models. A cell basically consists of three parts: inside (*intracellular*), boundary (the *membrane*) and outside (*extracellular*). Both the intracellular and extracellular media are fluids with many free ions (i.e. electrically charged particles). The cell membrane is selectively permeable, with varying degrees of permeability to some of these ions, and so an electrical potential difference across the cell membrane will be established, as some of these positively or negatively charged particles are allowed to move across the boundary. By convention, the extracellular fluid is electrical ground (i.e. it has a potential of 0 V) and the intracellular medium will typically have a potential (relative to extracellular) of up to ± 100 mV. The basic system is depicted in Fig. 5.1.

Hence, the cell is essentially an imperfect boundary separating two separate collections of chemicals such that these chemicals can pass across that boundary via various mechanisms. The net effect of these chemicals moving about is that, because of the electrical charges associated with the chemical species (these are ions and cations, rather than complete molecules), there is a net movement of charge across the cell boundary and the electrical balance between the inside and the outside of the cell is affected by this redistribution of chemical species.

Of course, the human body, along with every living organism, is made up of various different sorts of cells. For the sake of definiteness, we will focus on one particular type of cells: *neuron*. The brain consists of a very large (about 10^{12}) number of cells of a very large (about 10^3) number of different types, Nonetheless, the cells in the human brain fall into two basic types: *neuroglial* cells and *neurons*. Neuroglial cells (as far as we are concerned here) have no significant function. Despite making up about one-third of the cells in the brain there is no definitive description of their functional role. Neurons, on the other hand, are the stuff that makes us us. These are the

111

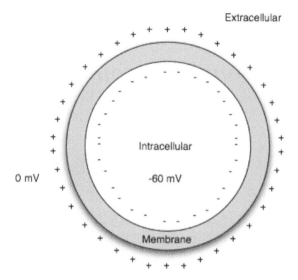

Figure 5.1:   A model cell. Basic caricature of the structure of a cell.

basic functional units which drive our memory and thought processes, and ultimately, somehow, lead to consciousness[1].

Traditional neurological science divided the brain into distinct regions. Each region is associated with a distinct functional role. However, modern imaging techniques now allow us to probe the activation of different regions of the brain during different functional activies and it is becoming clear that this picture is somewhat simplistic. The human brain consists of a vast array of different types of neurons with distinct functions and complex interconnectivity. Individually, these neurons are able to process information relatively slowly. Because we now know that neuronal information is encoded and computed based on individual spikes in electrical potential (and very probably the rate of transmission of these voltage spikes), the rate of information entering and leaving a neuron can be measured and a quantitative estimate of the neuronal information processing capability can be obtained: about 10–30 bits per second (bps). Any computational unit capable of processing even 100 bps is rather feeble. However, the human brain consists of 10^{12} such processing

---

[1]We suppose this to be the ultimate seat of human consciousness, but we are still a very long way from describing how this may be possible. While the focus of this text is on the mathematical and computational description of the dynamics of biological systems, it has been speculated [27] that quantum mechanical effects at a very tiny level (tiny on a biological scale, but perhaps not sufficiently tiny on a quantum mechanical level) may underpin the processes that eventually lead to consciousness. This hypothesis is appealing (and recent results have found evidence for quantum mechanical effects in botanical cellular systems [31]) because it puts human consciousness beyond the realm of computational simulation.

units working in parallel and each of these 10^{12} units is connected to around 10^3 other processors.

The combinatorial complexity of describing the human brain should now be readily evident. Even with this rather simplistic description — 10^{12} units of 10^3 different types connected to 10^3 other units — the numbers are fantastic. There are over

$$\binom{10^{12}}{10^3} \times \binom{10^{12}}{10^3} = \left(\frac{10^{12}!}{10^3! \times (10^{12} - 10^3)!} \right)^2$$

possible combinations — this is quite a big number. Nonetheless, our job here is not to describe the complete function of the human brain, only to highlight the difficulties of such a task. Rather, we are going to focus on a single neuron: one cell of 10^{12}. As Douglas Adams said: "the universe is big, really, really big". The number of possible ways to configure a human brain is even bigger (far bigger than the number of molecules in the known universe[2]).

Now, back to the individual cell. A cell is just a biochemical boundary separating two slightly different chemical soups: one on the inside (the *intracellular* region) and one on the outside (the *extracellular* region). The cell boundary itself is the most important and most interesting bit. It consists of a *phospholipid bilayer* — which is just a rather complicated way of saying there are two layers of special molecules. These molecules are fixed and locked together and therefore act as a wall. They have the rather useful[3] property that one end is *hydrophilic* and the other end is *hydrophobic*. That is, they have a water "loving" and a water "hating" end. One end of the molecule is attracted to water, one end is repelled by it (and attracted to oil). In the cell boundary these molecules arrange themselves in two layers with the hydrophilic ends pointing out and the hydrophobic ends pointing in. This functions exactly like double glazing[4]. The molecules align themselves in two layers, creating a boundary region between two watery regions. The boundary itself consists of an oily gap region[5]. Of course, in biology, nothing is so perfect as this description. In reality, the phospholipid bilayer is punctuated by various larger molecules, including such things as cholesterol and protein molecules, which allow for the transport of various charged chemical particles between the inside and the outside of the cell.

---

[2]Although not larger than the number of possible arrangements of these molecules — obviously!

[3]Useful to manufacturers of the class of food additive known as *emulsifiers*.

[4]I apologise to my students in Hong Kong, where double glazing is not particularly common — and therefore does not pose a useful analogy.

[5]Students of engineering, which I hope includes my students in Hong Kong, should recognise this description as directly analogous to an electrical capacitor — a fact we return to at some length later in this chapter.

5.2   Action Potentials and Ion Channels

The essence of cellular dynamics is this movement of charged particles. The human brain, along with most of the human body consists of water with various chemical salts dissolved in it. Salts, of which ordinary table salt is a good example, consist of positively and negatively charged atoms. In the case of table salt, it consists of equal portions of sodium (Na^+) and chlorine (Cl^-). In solid form these are bound together in a ratio of one to one: hence the chemical formula for table salt — NaCl. However, when dissolved in the chemical soup of the human body, they separate. A relative imbalance between the concentration of sodium and chorine will create an electrical charge (this is how batteries work). The cell boundary works by controlling this imbalance between the inside and outside of a cell to generate an electrical potential: the *action potential*. The various chemical atoms (either *anions* if they hold a negative charge or *cations* if their charge is postive) transit through the cell boundary via *ion channels* which are, in turn, constructed from the various chemical impurities in the phospholipid bilayer.

It is the generation and propagation of these action potentials that drive computation in the human brain.

Each neuron receives input from around 10^3 other neuronal cells via the cell's *dendrites*. These dendrites act as passive connections to other neurons in the brain. These inputs come in the form of electrical signalling — a potential difference between the intracellular and extracellular media. At the receiving cell, these signals are summed (some signals can have a negative contribution). If the total of these signals is sufficient, then the receptor cell will become excited and generate an action potential which will propagate down its *axon* to interact with other cells via the dendrites of those cells. In the next section, we will study the generation and propagation of these chemical and electrical signals with basic physical equations derived from the tendency for chemical species to propagate over population and electrical gradients. The process is illustrated in Fig. 5.2.

The typical resting potential (when population and electrical gradients are balanced) of the cell is about -60 mV. Then, when the *transmembrane potential* (i.e. the potential across the cell membrane), V_m, exceeds this (i.e. $V_m > -60$ mV), one says the cell is *depolarised*. Conversely, if V_m is below this resting level, then the cell is *hyperpolarised*. The action potential in Fig. 5.2 occurs because once the cell becomes sufficiently depolarised, it exhibits a large deviation from the rest state. On returning to rest, the cell needs to recover before it can fire again — during this period it hyperpolarises and gradually returns to its usual rest state. At this point the cell is ready to be depolarised and generate a new action potential. This process of having a large deviation follows a relatively small (but yet sufficiently large) perturbation, and the requirement that the cell then needs to rest to be ready to

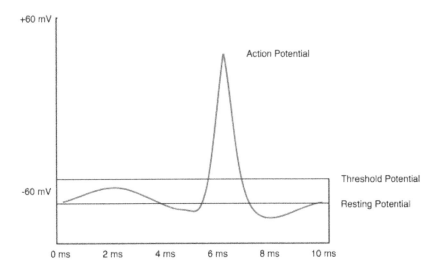

Figure 5.2: **The action potential.** In equilibrium, the interior and exterior of the cell remain in balance, at a potential of around −60 mV. When this equilibrium is disturbed (by the opening of *ion channels* through the cell membrane causing transport of ions) and the cell membrane reaches some threshold value, then an action potential results. That is, once the threshold potential is exceeded, the system will generate a large voltage spike followed by a return to equilibrium.

repeat this process is the defining characteristic of so-called *excitable media* — the subject of Chapter 8.

5.3 Fick's Law, Ohm's Law and the Einstein Relation

In essence, the various anions and cations inside and outside a cell will prefer to move to less densely populated regions — this is *Fick's law*. Conversely, *Ohm's law* tells us that like charges will tend to repel and seek electrical equilibrium. The *Einstein relation* defines a relationship between these two laws which can be used to balance the tendency to move toward less dense regions and the tendency to equalise charge.

Let's start with Fick's law. The flow of particles over a concentration gradient J_{diff} can be measured in units of concentration multiplied by velocity. Fick's law says that particles will diffuse from areas of high concentration to low concentration at a rate proportional to the concentration gradient. That is,

$$J_{diff} = -D\frac{dI}{dx}, \tag{5.1}$$

where I is the concentration and x is the spatial co-ordinate. Assuming that the cell is spherically symmetric, we assume one significant spatial direction and x is therefore the distance through the cell membrane. The constant D has units of m^2/s and measures the relative diffusivity of a particular system. This *diffusivity constant* could be influenced by physical constraints as well and thermodynamics (hotter systems diffuse more quickly).

The second relationship, Ohm's law, is familiar to all electronic and electrical engineering students — although in a slightly different form. Let J_{drift} denote the movement of charged particles under an applied voltage V. Ohm's law states that

$$J_{drift} = -\mu Z I \frac{dv}{dx}, \tag{5.2}$$

where Z is the ionic valence, and v is the applied voltage. As before, I is the ion concentration and x is the spatial distance parallel to the applied voltage (we assume that the cell is spherically symmetric and that this is still radial). The parameter μ (which has units of m^2/s) is known as the *mobility* and is a measure of the tendency for ions to move in a particular situation. The ionic valence Z is a dimensionless integer equal to the charge of a given ion. For example, for Cl^- $Z = -1$ and for Ca^{2+} $Z = +2$.

Finally, the Einstein relation rather surprisingly links diffusivity D in Eqn.

(5.1) and mobility of Eqn. 5.2

$$D = \frac{KT\mu}{q}, \tag{5.3}$$

where T is the temperature (in Kelvin) and K and q are physical constants. The constant $K = 1.3806 \times 10^{-23}$ J/K is the *Boltzmann constant* and $q = 1.60186 \times 10^{-19}$ C is the charge of an electron (in units of Coulombs).

In the next section we use these three relationships to establish equilibrium cellular membrane potential, and to determine the dynamical effect on the system of altering either permeability or ion concentration.

5.4 Cellular Equilibrium: Nernst and Goldman

Equations (5.1) and (5.2) describe two competing mechanisms for movement of ions through a cell membrane. Let's first consider a cell permeable only to one type of ion — potassium K^+. To determine the total flow we need to combine these

$$\begin{aligned}
J &= J_{diff} + J_{drift} \\
&= -D\frac{d[K^+]}{dx} - \mu Z[K^+]\frac{dv}{dx} \\
&= -\frac{KT}{q}\mu\frac{d[K^+]}{dx} - \mu Z[K^+]\frac{dv}{dx},
\end{aligned} \tag{5.4}$$

where $[K^+]$ denotes the concentration of potassium ions as a function of location x (radial displacement through the cell membrane). In equilibrium we have $J = 0$ and hence

$$\frac{KT}{q}\frac{d[K^+]}{dx} = -[K^+]\frac{dv}{dx}. \tag{5.5}$$

Integrating both sides with respect to x gives

$$\int \frac{KT}{q}\frac{1}{[K^+]}\frac{d[K^+]}{dx}dx = -\int \frac{dv}{dx}dx.$$

$$-\int_{v_o}^{v_i} dv = \frac{KT}{q}\frac{d[K^+]}{[K^+]}, \tag{5.6}$$

where we've made the substitution $Z = +1$. Finally, we obtain the *Nernst equilibrium potential* for a single ion species K^+,

$$\begin{aligned}
E_{K^+} &= v_i - v_o \\
&= \frac{KT}{q}\ln\frac{[K^+]_o}{[K^+]_i},
\end{aligned} \tag{5.7}$$

where the subscript i or o denotes the value of ion concentration of voltage on the interior or outside of the cell membrane. At room temperature $\frac{KT}{q} \approx 26$ mV and hence this value is commonly used in

$$E_{K+} \approx 26 \ln \frac{[K^+]_o}{[K^+]_i} \quad mV.$$

It is a trivial exercise to repeat this derivation and derive the Nernst potentials for the other common ion species:

$$E_{Na^+} = \frac{KT}{q} \ln \frac{[Na^+]_o}{[Na^+]_i} \tag{5.8}$$

$$E_{Cl^-} = \frac{KT}{q} \ln \frac{[Cl^-]_i}{[Cl^-]_o} \tag{5.9}$$

$$E_{Ca^{2+}} = \frac{KT}{2q} \ln \frac{[Ca^{2+}]_o}{[K^{2+}]_i}. \tag{5.10}$$

But, notice that the ratio is reversed for anions, and valence-two cations exhibit half the constant of proportionality.

Each of these relationships has been derived under rather unrealistic assumptions. When Eqns. (5.1) and (5.2) are set to balance, we assume that there is no other forces at work. That is, in the derivation of Eqn. (5.7) we assume that potassium is the only ion which can pass through the cell membrane. Of course, once we allow the possibility that the cell membrane may be permeable to other ions, the electrical balance is changed. Hence, the *Nernst potentials* of Eqns. (5.7 through 5.10) express the equilibrium voltage of a single ion specie. This does not help us to determine the overall equilibrium potential because a cell is typically permeable to a whole range of different ions (often, cells are most permeable to just three: potassium, sodium and chlorine).

Hence, to address this issue we turn to the so-called *Goldman equation*. Let us re-consider, first of all, just the flow of potassium J_{K+}:

$$J_{K+} = -\frac{KT}{q} \mu_{K+} \frac{d[K^+]}{dx} - \mu_{K+} Z[K^+] \frac{dv}{dx} \tag{5.11}$$

$$= -\frac{KT}{q} \mu_{K+} \frac{d[K^+]}{dx} - \mu_{K+} Z[K^+] \frac{V}{\delta}$$

$$= -P_{K+} \delta \mu_{K+} \frac{d[K^+]}{dx} - \frac{-P_{K+} q}{KT} V[K^+], \tag{5.12}$$

where Eqn. (5.11) is exactly the expression we used in Eqn. (5.4) to obtain the Nernst potential (by setting $J_{K+} = 0$). But now we do not assume that the overall flow is zero. The ratio $\frac{V}{\delta}$ is the potential difference through the cell membrane over the thickness of that membrane (assuming that the voltage field v changes at a constant rate and so $\frac{dv}{dx} = \frac{V}{\delta}$). The quantity P_{K+} is the permeability of the cell membrane to potassium and is just an abbreviation of the various constants in the equation

$$P_{K+} = \frac{\mu_{K+}KT}{\delta q} \tag{5.13}$$

$$= \frac{D_{K+}}{\delta}.$$

Rearranging and then integrating (in a manner similar to what was needed for the Nernst potential), we can obtain

$$\int_0^\delta dx = -\int_{[K^+]_i}^{[K^+]_i} \frac{d[K^+]}{\frac{J_{K+}}{P_{K+}\delta} + \frac{qV[K^+]_o}{KT\delta}}$$

$$\delta = -\frac{KT\delta}{qV} \ln \left(\frac{\frac{J_{K+}}{P_{K+}\delta} + \frac{qV[K^+]_o}{KT\delta}}{\frac{J_{K+}}{P_{K+}\delta} + \frac{qV[K^+]_i}{KT\delta}} \right)$$

$$-\frac{qV}{KT} = \ln \left(\frac{\frac{J_{K+}}{P_{K+}\delta} + \frac{qV[K^+]_o}{KT\delta}}{\frac{J_{K+}}{P_{K+}\delta} + \frac{qV[K^+]_i}{KT\delta}} \right). \tag{5.14}$$

Then, after some algebraic reshuffling we obtain from Eqn. (5.14) the following identity for potassium

$$J_{K+} = \frac{qVP_{K+}}{KT} \left(\frac{[K^+]_o - [K^+]_i e^{-\frac{qV}{KT}}}{e^{-\frac{qV}{KT}} - 1} \right). \tag{5.15}$$

By the same reasoning we can also obtain an equivalent expression for chlorine:

$$J_{Cl-} = \frac{qVP_{Cl-}}{KT} \left(\frac{[Cl^-]_i - [Cl^-]_o e^{-\frac{qV}{KT}}}{e^{-\frac{qV}{KT}} - 1} \right), \tag{5.16}$$

where the sign and the position of interior and exterior concentrations are reversed for the cation.

Finally, if we assume that the cell membrane is permeable only to potassium and chlorine, we can combine Eqns. (5.15) and (5.16), subject to the assumption of space-charge neutrality $J_{K+} = J_{Cl-}$, and obtain the two species Goldman equation

$$V_m = \frac{KT}{q} \ln \left(\frac{P_{K+}[K^+]_o + P_{Cl-}[Cl^-]_i}{P_{K+}[K^+]_i + P_{Cl-}[Cl^-]_o} \right). \tag{5.17}$$

Of course, an expression similar to Eqns. (5.15) and (5.16) can be derived for a third ion specie (the third most permeable ion for many cells, after potassium and chlorine, is sodium),

$$J_{Na+} = \frac{qVP_{Na+}}{KT} \left(\frac{[Na^+]_o - [Na^+]_i e^{-\frac{qV}{KT}}}{e^{-\frac{qV}{KT}} - 1} \right). \tag{5.18}$$

So now the assumption of space charge neutrality means that $J_{K^+} + J_{Na^+} = J_{Cl^-}$ and hence, for these three species, the equilibrium membrane potential is

$$V_m = \frac{KT}{q} \ln \left(\frac{P_{K^+}[K^+]_o + P_{Na^+}[Na^+]_o + P_{Cl^-}[Cl^-]_i}{P_{K^+}[K^+]_i + P_{Na^+}[Na^+]_i + P_{Cl^-}[Cl^-]_o} \right). \quad (5.19)$$

Of course, one could continue adding additional species and obtain ever more complex expressions of the form of Eqns. (5.19) and (5.17). However, it is often the case that, after the first three species, the permeabilities of additional species are relatively small (the P_X terms approach zero) and so either Eqn. (5.19) or (5.17) is an adequate approximation.

So far in this section we have been concerned with equilibrium potentials: either the equilibrium when a cell is permeable to only one specie (Nernst potentials) or when multiple species interact (Goldman equilibrium). However, this does not tell us much about the dynamics of a single cell (and it is, after all, dynamics which we are interested in). In the next section we will use this information about equilibria to develop models of the dynamical behaviour of cells. From these models (which initially will be electronic circuits) we will eventually develop a mathematical description of the behaviour of various classes of neurons (in Chapter 6).

5.5 Equivalent Circuits

In Chapter 2 we analysed mathematical descriptions of biological dynamics by first describing the equilibria of the system — the fixed points. In the previous section we obtained expressions for the equilibrium cell membrane potential under various conditions. That is, the Nernst potentials and Goldman equations of the previous section quantify various steady states of the cell. In this section we introduce dynamical behaviour by building an electronic circuit model of the cell membrane.

Let us first consider the effect of the individual ion channels. Suppose that a cell membrane is permeable to one particular type of ion. That ion will have associated with it some particle equilibrium potential V_{ion}, which, when this ion is considered in isolation from everything else, is nothing more than the corresponding Nernst potential. However, the ions passing through the cell membrane do not do so for "free": there is some cost in terms of resistance to the physical passage of the ions. The ions themselves are charged particles (either positively or negatively) and so their movement generates an electric current. The physical resistance to their movement through the cell wall creates a loss of electrical potential (that is, a voltage drop) and so can be modelled exactly as an electrical resistance. The situation is depicted in Fig. 5.3.

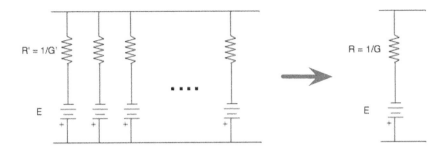

Figure 5.3: Ion channels. The cell membrane is modelled as being permeable to a particular ion specie. There is a certain resistance R (per unit cell membrane surface area) and electric potential E associated with the transport of ions across the boundary. Over a larger surface area this can be treated as a sequence of resistances in parallel and modelled as a single resistance and voltage source (on the right).

Of course, with a larger area of cell membrane there is more opportunity for ions to pass through the barrier. The entire cell membrane can therefore be modelled as a collection of resistances in parallel. This reduces to a single resistance and voltage source, as shown on the right of Fig. 5.3.

With multiple different ion species, the situation is only marginally complicated. Each individual ion specie obeys the relationship depicted in Fig. 5.3. Hence over a section of cell membrane, there is a resistance on potential for each ion specie: R_{NA} and $_{Na}$ for sodium, R_K and $_K$ for potassium, and R_{Cl} and $_{Cl}$ for chlorine. This is illustrated in Fig. 5.4.

We are now in a position to apply elementary circuit analysis to the model in Fig. 5.4. Although this model is incomplete (below we will account for the effect of active ion transport and membrane capacitance), it is instructive to first ask what would happen with only this model.

Refer to Fig. 5.4. Let i_1 be the current circulating in the left-hand loop and i_2 be the current in the right half loop (both clockwise). Hence the current through the sodium channel is i_1 (outwards), through the potassium channel is $i_2 - i_1$ and through the chlorine channel is $-i_2$ (again taking the outwards convention). Hence, we have

$$-v_m = E_{Na} + R_{Na}i_1 \qquad (5.20)$$
$$= E_K + R_K(i_2 - i_1) \qquad (5.21)$$
$$= E_{Cl} - R_{Cl}i_2. \qquad (5.22)$$

Rearranging the first and third we obtain expressions for i_1 and i_2 which can then be substituted into the middle equation (with some small amount of

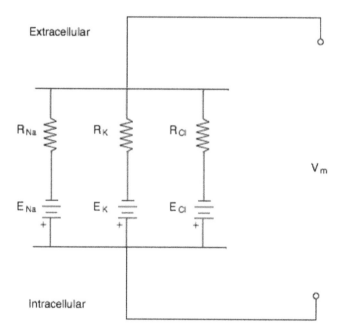

Figure 5.4: Na, K and Cl ion channels. The model of Fig. 5.3 is repeated with a separate resistance and voltage source for each ion specie. Of course, the arrangement of these in parallel could then again be reduced to a single resistance and potential which would summarise the effect of all species.

algebraically hard labour):

$$V_m = \frac{R_{\text{Cl}}R_{\text{K}}E_{\text{Na}} - R_{\text{Na}}R_{\text{Cl}}E_{\text{K}} - R_{\text{K}}R_{\text{Na}}E_{\text{Cl}}}{R_{\text{Na}}R_{\text{Cl}} + R_{\text{K}}R_{\text{Cl}} - R_{\text{Na}}R_{\text{K}}}, \tag{5.23}$$

which of course is what one derives from a Thévenin equivalent circuit (which would be familiar to those with a background in circuit theory)[6]. Briefly this gives a membrane potential which is somehow a weight average of the passive membrane potential due to the Nernst potentials of the individual species. In fact this is equivalent to the Goldman equation, except that here the expression is in terms of membrane electrical resistances and the Goldman equation is derived for known relative permeabilities. The permeabilities and the resistances are not trivially related.

In what we have described so far, the movement of ions is *passive*. That is, ions move through the cell boundary only due to diffusion (from high density to low density) and electromotive force (in an attempt to equalise the charge). However, in most cells there are also various *active transport* mechanisms. *Ion pumps*, for example, actively force ions of a particular specie across the cell boundary at a particular rate. This essentially creates a current source across the cell boundary in Fig. 5.4: the ions being forced across the cell boundary all carry a charge and so their forced movement induces a current. These ion pumps exist in many different forms: they may transport a single specie (for example, a calcium ion pump would transport only calcium) or they may transport particular species in a particular ratio.

The sodium-potassium ion pump, for example, transports a fixed number of sodium and potassium ions across the cell membrane in particular directions. For example, in one particular form, this ion pump transports three Na^+ ions out through the cell membrane in exchange for two K^+ potassium ions coming in. The operation of this pump may seem at first to be conceptually somewhat mysterious. It is easy to think about it as a biochemical gate in the cell membrane. When two potassium ions are present on the outside (and engaged at this gate) and three sodium ions are engaged on the inside of the gate, then the gate opens and physically exchanges the ions between outside and inside. Never mind the biochemical function, the electronic circuit model is straightforward. In Fig. 5.5 we add the current source for both the transport of sodium and potassium.

Finally, we now consider one last property of the cell membrane, which was hinted at in Section 5.1, and this will then give us a sufficiently complete model of cellular dynamics. Recall that the cell boundary consists of a phospholipid bilayer: an oily boundary between the watery interior and exterior. This

---

[6]For those non-electronic engineers, the *Thévenin equivalence theorem* provides a method of transforming a more complex circuit into a simpler circuit composed of fewer elements with different parameter values. In this case, for example, we can transform multiple pairs of resistors and voltage sources into a single resistor and a single voltage source. This is actually illustrated in Fig. 5.3.

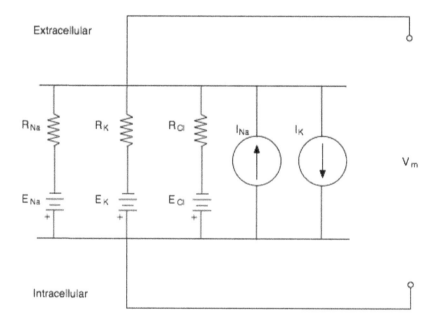

Figure 5.5: Na-K ion pump. The sodium and potassium ion pumps model the active transport of ions across the cell boundary and generate the corresponding current sources. In this case, since the transport of sodium and potassium ions are locked in a ratio of 3:2, we also have the restriction that $2I_{Na} = 3I_K$. It is therefore trivial to replace these two ion pumps with a single outward current source with a current of $\frac{1}{2}I_{Na}$.

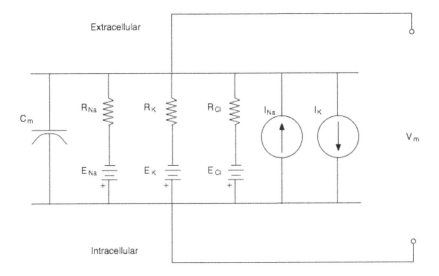

Figure 5.6: **Cell membrane.** The cell membrane itself acts exactly analogously to an electrical capacitor — the relatively slow leakage of current across the cell boundary due to the presence of the phospholipid bilayer creates an electrical memory. To model this effect, it is only necessary to add a capacitor, as shown above.

effectively creates a natural capacitor — exactly analogous to the standard circuit element. In electronics, a capacitor is constructed from two parallel charged plates separated by a partially insulated electrolyte. In the cell, the two layers of phospholipids (which are partially permeable to charged ions) form the charged plates and the oily layer is an electrolytic boundary. Hence, to model this effect, we need only add a capacitor to the previous circuit diagram and obtain the picture presented in Fig. 5.6.

5.6   Dendrites

We now consider a slightly different, but also important cellular model. In Section 5.5 we built a circuit model for the dynamic interaction between the inside and outside of a cell: that is, dynamics across the cell boundary. However, it is also important to understand how electrical signals are propagated along tubular cell membranes — along *dendrites*. Remember that dendrites exist to propagate electrical signals from one cell to another. The important dynamical behaviour is therefore how the electrical signal can pass along the tube (rather than in the previous section where we were concerned with transport across the boundary).

In Fig. 5.7 we give a simple circuit model for a section of dendrite. Essentially, we consider small sections of a long tube. Within each small dendritic section we have a simple resistor-capacitor-voltage source model of the membrane dynamics — exactly as derived in Section 5.5. That is, there is some capacitive effect C_m, some resistive loss along the length of the dendrite R_a and some Thévenin equivalent model for the various (passive) ion channels V_{Th} and R_{Th}.

If a current is injected into the left-hand side of the model (that is, at one end of the dendrite), what is the voltage observed at the other end? As drawn in Fig. 5.7, the observed voltage varies with location along the dendritic length (observe that the value v_m depends on the location of the lower probe relative to the sequence of resistors R_s).

We can obtain a useful approximation with a rather simple-minded approximation. Suppose that the system is in a steady state and that the current injection is a constant (DC[7]): we may then replace the capacitance C_m with open circuits. If we also neglect the Thévnin voltages sources as being relatively small, we arrive at the simplified revised circuit illustrated in Fig. 5.8.

---

[7]That is, *direct current*. DC circuits have the useful (from the point of view of circuit analysis) property that capacitive circuit elements appear as open circuits.

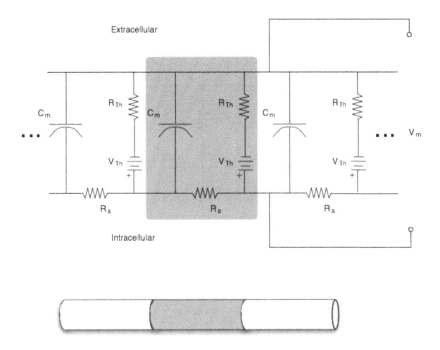

Figure 5.7: **Dendrites.** The resistance R_a is the resistance to flow of charge along the length of the dendrite. The resistance R_{Th} and voltage V_{Th} account for the various ion channels leaking current through the cell walls, and the capacitance C_m is the cellular membrane capacitance.

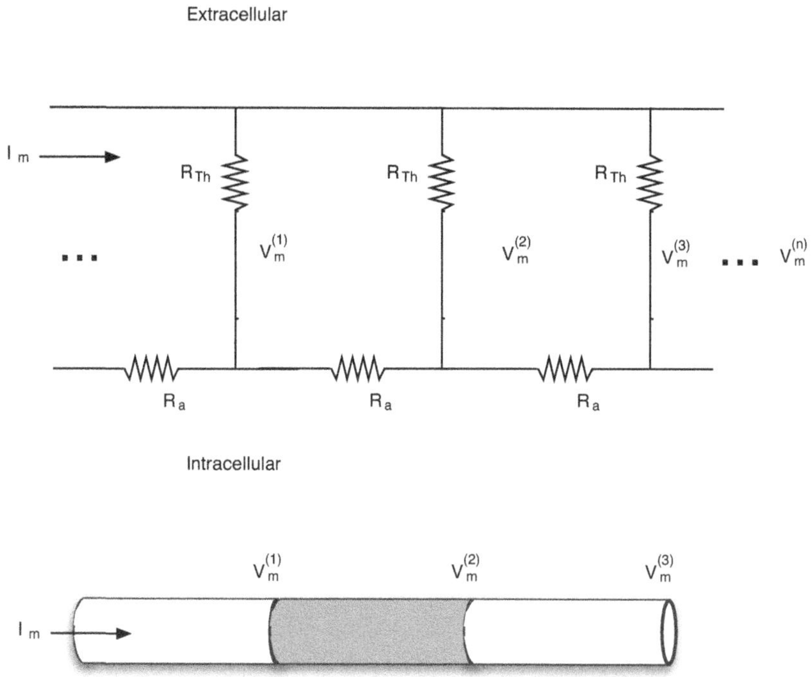

Figure 5.8:   **Dendrites.** A simplification of Fig. 5.7.

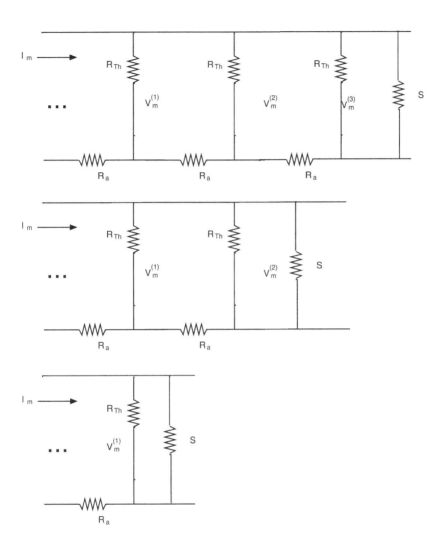

Figure 5.9: Dendrites. Assuming that the dendritic length n is relatively long, then the total load S will be the same over most segments i, provided $i \ll n$. Hence, we can simplify Fig. 5.8 further and obtain an expression for the total resistance.

We can now consider the total current in each path. Assuming a very large number of segments, then most of the current will path through a resistance $R_s + R_{Th}$ (corresponding to the left-most resistors in Fig. 5.8). A correspondingly smaller portion of the current will pass through the next link and incur a resistance of $2R_s + R_{Th}$. After n links the resistance will be $nR_s + R_{Th}$. The question, of course, is how much current is seen over the last resistor in this chain and therefore must incur a total resistance of $nR_s + R_{Th}$. Suppose that the whole circuit (as seen from the left in Fig. 5.8) is given by S. Now for a very large length n, this is also the resistance which would be seen after discounting the first resistances R_{Th} and R_a. That is, the three circuits in Fig. 5.9 are equivalent to one another.

Remember that if two resistances R_1 and R_2 are in series, the total effect is $R_1 + R_2$. If two resistances are in parallel, then their total effect is $\frac{1}{\frac{1}{R_1}+\frac{1}{R_2}}$. Hence the total resistance S of the cable is given by (referring to Fig. 5.9)

$$S = R_a + \frac{1}{\frac{1}{R_{Th}} + \frac{1}{S}}$$

$$0 = S^2 R_a - R_a R_{Th}$$

$$S = R_a + \sqrt{R_a + 4R_a R_{Th}} \qquad (5.24)$$

where Eqn. (5.24) is the positive root of the quadratic (since the resistance must be positive).

Applying the same symmetry arguments, we can also ask how much the dendritic voltage drop by over successive segments

$$v_m^{(1)} = \frac{S - R_a}{S} I_m$$

$$v_m^{(2)} = \frac{S - R_a}{S + R_{Th}} \frac{R_{Th}}{S} I_m$$

$$\frac{v_m^{(2)}}{v_m^{(1)}} = \frac{R_{Th}}{S + R_{Th}} \qquad (5.25)$$

$$\frac{v_m^{(1)} - v_m^{(2)}}{v_m^{(1)}} = \frac{S}{S + R_{Th}}. \qquad (5.26)$$

Alternatively we can take a continuous approach and employ nineteenth-century cable theory which was developed to describe the transmission of electrical power over power lines. Rather than consider a series of discrete chunks of dendrite with characteristic resistances R_{rmTh} and R_a, we now consider R_a and R_{rmTh} to be resistance per unit length. The complete cable equation $\tau \frac{\partial V}{\partial t} = \lambda^2 \frac{\partial^2 V}{\partial x^2} - V$ has a characteristic length scale $\lambda = \sqrt{\frac{R_{Th}}{R_a}}$ as well as a time scale $\tau = R_{Th} C_m$. But since we have already restricted ourselves to the steady-state solution, we can let $C_m = 0$ and then we have the system

$$V = \frac{1}{\lambda^2} \frac{\partial^2 V}{\partial x^2}, \qquad (5.27)$$

which immediately yields the solution

$$V(x) = V(0)e^{-\frac{x}{\lambda}}, \tag{5.28}$$

where $V(0)$ is the voltage at injection (i.e. subject to the full load S, and hence $V(0) = I_m S$) and $V(x)$ is the potential at some point x further along the dendrite (of course, the length units of x, R_a and R_{rmTh} must be the same).

5.7 Summary

In this chapter we have developed models of cellular dynamics that can explain the simple molecular balance between the interior and the exterior of a cell. Most of this description was in the context of a steady state — that is, we examined the flow and search for a stable equilibrium. In the next chapter we develop this idea further. Based on the concept that ion channels can open and close to allow ingress and egress of charged particles, and that these dynamics can occur with different time scales, we develop differential equation descriptions for a large range of cellular dynamics. We will use these descriptions, and the analysis tools we have developed in previous chapters, to obtain a detailed description of the type of dynamical changes that neurons can undergo — and the corresponding range of behaviour that they can exhibit.

Glossary

Action potential The electrical pulse signal resulting from the excitation of a neuron.

Active transport The movement of ions (of a particular specie) over an ion channel with external forcing. In effect, the external forcing creates a motive force moving ions through the barrier at a particular rate.

Anion A negatively charged chemical ion.

Axon The cable-like extension from a neuron allowing electrical signals to be conducted from the cell.

Cation A positively charged chemical ion.

Cellular membrane The barrier between the inside and the outside of a cell. The cellular membrane consists of a phospholipid bilayer which forms a capacitive semi-permeable barrier.

Dendrite The branching extensions from a neuron which allow the conduction of electrical signals into a cell.

Depolarised The state of a cell when its transmembrane potential becomes large (exceeding, say, -60 mV).

Einstein relation A mathematical relationship linking diffusitivity and mobility (and hence Fick's and Ohm's laws).

Extracellular Outside the cell.

Fick's law A mathematical expression for the flow of particles in the presence of a concentration gradient (particles prefer areas of low concentration).

Goldman equation The extension of the Nernst potential to the case with multiple different ion species each with a different relative permeability (through the cell membrane) in the same model.

Hyperpolarised The state of a cell when its transmembrane potential becomes small (significantly less than, say, -60 mV).

Intracellular Inside the cell.

Ion channel A pathway through the cellular membrane over which certain ions (and only those ions) may pass.

Ion pump An active transport ion channel.

Neuroglial A type of cell, thought to be responsible for maintaining chemical equilibrium in the brain (and not information processing).

Neuron A type of cell. Responsible for information processing in the brain.

Nernst potential The value of transmembrane potential at which the flows of a single specie of ion due to concentration and electrical gradients are balanced.

Ohm's law A mathematical expression for the flow of charged particles in the presence of an electric field.

Passive transport The movement of ions (of a particular specie) over an ion channel with no external forcing. That is, the ion channel essentially acts as an open doorway for a particular specie.

Phospholipid bilayer A pair of layers of phospholipids. A phospholipid is a molecule with a water-loving and a water-hating end. The cells are arranged in a layer so that the water-loving ends and the water-hating ends are aligned. Hence, one side of the layer is attracted to water, the other not.

Thévenin equivalence theorem A mathematical theorem providing a representation of an electrical circuit in a specific canonical form.

Transmembrane potential The electrical potential difference across the cellular membrane.

Exercises

1. What is an action potential? Where does it come from? Draw one, correctly labelling regions of depolarisation and hyperpolarisation.

2. Extend the Goldman equation from two species to three (K^+, Na^+ and Cl^-). Confirm the result given in lectures.

3. A cell in equilibrium is populated internally with 300 mM of K^+ ions and externally with 150 mM. What is the potassium Nernst potential? Suppose that the same cell is also populated with 500 mM of Cl^-. At electric equilibrium, what is the external concentration of Cl^-?

4. Consider a neuron permeable to sodium, potassium and chlorine. Given the rate of permeability and concentration of each species (in the following table), compute the Nernst potential for each ion and the membrane potential V_m at normal body temperature (37°C). Comment on the observed discrepancy between individual Nernst potentials and the actual membrane potential.

Ion	Cytoplasm (mM)	Extracellular (mM)	Relative Permeability	Conductance (μS)
K^+	250	15	1	2.6
Na^-	30	280	0.06	48
Cl^-	40	350	0.35	8.2

5. A cell permeable to sodium, potassium and chlorine ions has ion concentrations and permeabilities as given in the following table.

Ion	Cytoplasm (mM)	Extracellular (mM)	Relative Permeability	Conductance (μS)
K^+	200	20	1	2
Na^+	30	240	0.05	50
Cl^-	50	350	0.35	10

(a) Compute the resting cellular membrane potential (T = 310K).

(b) Compute the change in potential due to an inward current pulse of amplitude I_a and duration τ.

6. Explain why a voltage source, rather than a current source, is used in the equivalent circuit models for ion exchange through the cell membrane.

7. Ion concentrations in squid axoplasm and blood have the following values (in mM):

Ion	Axoplasm	Blood
K^+	400	20
Na^+	50	440
Ca^{2+}	0.4	10
Mg^{2+}	10	54
Cl^-	123	560

(a) Compute the Nernst potential of each specie and draw the equivalent circuit model.

(b) Solve the equivalent circuit and deduce the membrane potential.

(c) Explain the assumptions underlying these results. That is, what have you actually computed?

8. A cell permeable to sodium, potassium and chlorine ions has ion concentrations and permeabilities as given in the following table:

Ion	Cytoplasm (mM)	Extracellular (mM)	Relative Permeability	Conductance (μS)
K^+	150	20	1	2
Na^+	45	240	0.05	60
Cl^-	40	300	0.35	20

(a) Compute the Nernst potential for each specie (T = 310K).

(b) Draw, label and fully describe the equivalent circuit.

(c) Compute the resting cellular membrane potential.

9. Ion concentrations in squid axoplasm and blood have the following values (in mM):

Ion	Axoplasm	Blood
K^+	350	25
Na^+	30	400
Ca^{2+}	0.35	12
Mg^{2+}	15	45
Cl^-	123	560

(a) Compute the Nernst potential of each specie and draw the equivalent circuit model.

(b) Solve the equivalent circuit and deduce the membrane potential.

(c) Explain the assumptions underlying this results. That is, what have you actually computed?

10. Show that Ohm's law (5.2) is equivalent to the relation $V = IR$, which you are (possibly) more familiar with.

11. Ion concentrations in a red blood cell have the following values (in mM):

Ion	Axoplasm	Blood	Relative Permeability
K^+	125	5	1
Na^+	10	150	0.55
Cl^-	80	115	0.21
Mg^{2+}	5	1.2	0
Ca^{2+}	0.35	15	0

(a) Compute the Nernst potential of each specie.

(b) Compute the membrane potential.

(c) Draw an equivalent circuit, labelling the appropriate values.

(d) Explain the difference between the potentials obtained in (a) and (b).

(e) What is the meaning of the zero values for relative permeability?

12. A (frog muscle) cell permeable to sodium, potassium, calcium and chlorine ions has ion concentrations and permeabilities as given in the following table:

Ion Ion	Cytoplasm (mM)	Extracellular (mM)	Relative Permeability	Conductance (µS)
K^+	125	2.25	0.8	65
Na^+	10.5	110	1	80
Ca^{2+}	10^{-4}	2	0.05	2
Cl^-	1.5	75	0.30	5

(a) Compute the electrical resistance for each ion.

(b) Compute the Nernst potential for each specie at 37°C.

(c) Suppose that the system also exhibits a Na^+-K^+ ion pump and the membrane has a capacitance C_m. Draw, label and describe the equivalent circuit. Where possible, give numerical values of resistance and voltage. Do not perform any additional calculations.

(d) Suppose that a certain pharmaceutical agent causes the relative permeability of this cell to calcium to drop to 0. Under this situation, compute the equilibrium membrane potential.

13. Suppose that the neuron in Exercise 14 (above) also exhibits active sodium and potassium ion pumps. Construct a circuit model for this situation and derive a relationship between the current in these pumps and the membrane potential. Assume that the capacitance of this cellular membrane is 2 pF. Finally, compute the current required by the ion pumps to maintain a membrane potential of 60 mV.

14. The circuit depicted in Fig. 5.10 is a model of voltage potential across a cell membrane.

(a) Using Kirchoff's laws, derive a system of two differential equations for voltage across the cell membrane $V_m = v_i - v_e$ in terms of an externally applied current.

(b) Using suitable transformation, show that this model is equivalent to the FitzHugh–Nagumo model. What is the usual form of $F(V)$?

(c) Draw nullclines and list the different possible dynamics that may occur in terms of these nullclines.

15. Assume a cell membrane is influenced by passive transport of potassium and sodium. Construct the equivalent circuit and therefore determine V_m. You may assume the following (experimentally determined) values for conductance and membrane potentials: $R_K = 1.7$ kΩ, $R_{Na} = 15.67$ kΩ, $E_K = -105$ mV and $E_{Na} = 56$ mV.

16. The cytoplasmic and extracellular ion concentrations for three species of ion in a squid giant axon at 6.3°C are given in the following table. Relative permeabilities of the four ion species are also given.

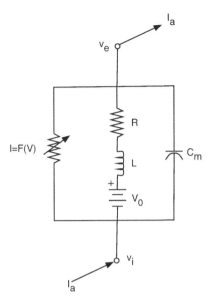

Figure 5.10: **Circuit diagram for Exercise 14.**

Ion	Cytoplasm (mM)	Extracellular Concentration (mM)	Relative Permeability
K^+	300	30	1
Na^+	45	200	0.1
Cl^-	70	450	0.6

(a) Compute the Nernst potential for each specie.

(b) Compute the membrane potential.

Chapter 6

Mathematical Neurodynamics

6.1 Hodgkin, Huxley and the Squid Giant Axon

In 1952, Alan Hodgkin and Andrew Huxley, two British physiologists, described a series of experiments conducted on a squid giant axon . The aim of these experiments was to develop a better understanding of the origin of action potentials. Eleven years later, their work led to the pair being awarded the Nobel Prize for physiology or medicine. Today the model stands as the foundation of the quantitative description of cellular dynamics. Initially they had worked with frog cells, but later moved to the squid giant axon (so named because it is a large axon in a normal-size squid — unlike the giant squid of Fig. 6.1) because this particularly large cell allowed them to apply the electrophysiological techniques available at the time to measure ionic currents — something which would not have been possible with smaller cells.

The mathematical description of cellular dynamics put forward by Hodgkin and Huxley is made all the more remarkable because of the way in which their work is the culmination of excellent experimental neurophysiology as well as brilliant mathematical insight and computational analysis. The computational analysis required integration with a mechanical calculator of the system of differential equations which was insolvable with analytic tools (see Fig. 6.2). Remarkably, although the digital computers of today make the task easier, it is still a highly nontrivial system. Here, we will provide a rather simplified description.

As we did in the previous chapter, we start with a simple model of cellular membrane potential V,

$$C_m \frac{dV}{dt} + I_{\text{ion}}(V, t) = 0, \qquad (6.1)$$

where C_m is the membrane capacitance (induced by the phospholipid bilayer), the membrane potential $V = V_{\text{inside}} - V_{\text{outside}}$ is the difference between the potential inside and outside the cell, and I_{ion} is some function of V and t which described the various ionic currents. Of course, the magnitude of the current due to some ions is significantly larger than others, and, just as in Chapter 5.1, the sodium and potassium currents are typically the most dominant. Hence,

Figure 6.1: **A giant squid.** Not to be confused with a squid giant axon. (Image obtained from http://commons.wikimedia.org/ and freely distributed in the public domain.)

Figure 6.2:   **The calculator.** A mechanical calculator of the type used by Andrew Huxley to intergrate the Hodgkin–Huxley equations. This particular machine comes from the office of Alan Hodgkin. (Image reproduced with permission from Hugh Robinson of University of Cambridge, Department of Physiology, Development and Neuroscience).

Eqn. (6.1) can be expanded as follows:

$$C_m \frac{dV}{dt} = -g_{\text{Na}}(V - V_{\text{Na}}) - g_{\text{K}}(V - V_{\text{K}}) - G_{\text{L}}(V - V_{\text{L}}) + I_{\text{app}} \quad (6.2)$$
$$= -g_{\text{eff}}(V - V_{\text{eff}}) + I_{\text{app}},$$

where

$$g_{\text{eff}} = g_{\text{Na}} + g_{\text{K}} + g_{\text{L}}$$
$$V_{\text{eff}} = \frac{(g_{\text{Na}} + g_{\text{K}}V_{\text{K}} + g_{\text{L}}V_{\text{L}}}{g_{\text{eff}}}.$$

The terms g_{Na}, g_{Na} and g_{L} are the *conductances* through the cell membrane for sodium ions, potassium ions and everything else (so-called *leakage* conductance). The corresponding V_{Na}, V_{K} and V_{L} are the single-specie equilibrium potentials: that is, the cellular equilibrium in the presence of only that specie. Finally, I_{app} is the applied current: this is the input signal received from other cells in the system. By combining these equations again, we obtain an effective conductance and voltage for the entire system: g_{eff} and V_{eff}. Hence, V_{eff} is the overall equilibrium voltage: $V_{\text{eq}} = V_{\text{eff}}$.

The clever experimental part of Hodgkin and Huxley's work was to then provide a more detailed description — based on experimental data – of the

mathematical behaviour of each of these conductance terms. They trans-
formed Eqn. (6.2) into the following system:

$$C_m \frac{dV}{dt} = -\hat{g}_{Na}(V - V_{Na} - \hat{g}_K(V - V_K - \hat{g}_L(V - V_L) + I_{app} \tag{6.3}$$

$$\frac{dm}{dt} = \alpha_m(1 - m) - \beta_m m \tag{6.4}$$

$$\frac{dn}{dt} = \alpha_n(1 - n) - \beta_n n \tag{6.5}$$

$$\frac{dh}{dt} = \alpha_h(1 - h) - \beta_h h. \tag{6.6}$$

In this system the three new dynamical variables m, n and h each describe the
different rates of flow of various ions of the cellular membrane. The various
parameters $\alpha_{m,n,h}$ and $\beta_{m,n,h}$ all have values which depend on the voltage V
and were all determined experimentally by Hodgkin and Huxley:

$$\alpha_m = 0.1 \frac{25 - v}{\exp\left(\frac{25-v}{10}\right) - 1}$$

$$\beta_m = 4 \exp\left(-\frac{v}{18}\right)$$

$$\alpha_n = 0.07 \exp\left(-\frac{v}{20}\right)$$

$$\beta_n = \frac{1}{\exp\left(\frac{30-v}{10}\right) + 1}$$

$$\alpha_h = 0.01 \frac{10 - v}{\exp\left(\frac{10-v}{10}\right) - 1}$$

$$\beta_h = 0.125 \exp\left(-\frac{v}{80}\right)$$

$$v = V - V_{eq}$$

$$\hat{g}_{Na} = 120$$

$$\hat{g}_K = 36$$

$$\hat{g}_L = 0.3$$

$$v_{Na} = 115$$

$$v_K = 12$$

$$v_L = 10.6.$$

However, for our purposes, this system, although neat, is still far more
complex than we can feasibly deal with. A simplified circuit model is presented
in Fig. 6.3, but the presence of variable resistances (describing the dynamics
of the various conductance terms $g_{Na} = \hat{g}_{Na}m^3 h$ and $g_K = \hat{g}_K n^4$) means that
it is still too complex to be analysed in detail. Nonetheless, we can glean
some slightly deeper insight into this system by taking a little closer look at
the dynamics of the conductance terms.

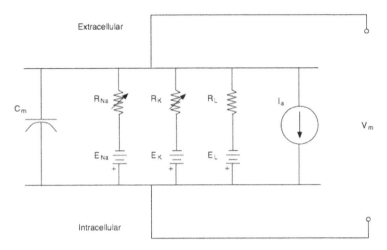

Figure 6.3: **An equivalent circuit model of the Hodgkin–Huxley equations.** The resistance $R_{\mathrm{Na}} = \frac{1}{g_{\mathrm{Na}}}$, $R_{\mathrm{K}} = \frac{1}{g_{\mathrm{K}}}$ and $R_{\mathrm{L}} = \frac{1}{g_{\mathrm{L}}}$. The current I_a is the applied current and the voltage sources $E_{\mathrm{Na,K,L}}$ are the membrane voltages.

Note that both g_{K} and g_{Na} are fourth order polynomial powers of similar variables: either n^4 or $m^3 h$. Let us then take a toy conductance $g = \hat{g}k^4$ where $\frac{dk}{dt} = \alpha_k(1 - k) - \beta_k k$. We can suppose that α_k and β_k are constants of the system (that they do not depend on any voltage V). This is reasonable because each of the variables m, n and h in the real model behave at different time scales and we are (for now) interested in studying the steady state behaviour. Then the solution to this differential equation is easily verifiable[1] $k(t) = \frac{\alpha}{\alpha + \beta} + \exp\left(-(\alpha + \beta)t\right)$. The function $k(t)$ is monotonically decreasing (if $\alpha - \beta$ is positive) and has asymptotic value

$$\lim_{t \to \infty} k(t) = \frac{\alpha}{\alpha + \beta}.$$

Hence, the parameter k converges exponentially to some equilibrium which is expressed as a function of $\alpha + \beta$. Hodgkin and Huxley had observed that the conductances behave like functions of some general form

$$(A + Be^{-Ct})^4,$$

and so this is precisely what they constructed with the variables m, n and h, albeit as functions of $\alpha_{\mathrm{m,n,h}}$ and $\beta_{\mathrm{m,n,h}}$.

All the variables in these equations have dimensional units: they are derived directly from measurement of a physical cell. The voltages are in mV, current

---

[1] Check this if you do not believe me.

density is in $\mu A/cm^2$, conductance in mS/cm^2 and capacitance in $\mu F/cm^2$. Notice, that the electrical parameters for current, conductance and capacitance are expressed as densities — per unit of surface area. That is, their actual magnitude depends on the actual surface area of the cell boundary under consideration.

In the next section we will make some simplifications to this model, so that we can analyse it to a more satisfactory depth. To do this, we first need to abstract the physically real system to a mathematical one with non-dimensional units.

6.2   FitzHugh–Nagumo Model

The Hodgkin–Huxley model (6.3) is a model of a squid giant axon which is taken as a biological model of more general neuronal cell dynamics. However, this model is rather complex. In this section we distill a still simpler model, the FitzHugh–Nagumo model, which is a model of the Hodgkin–Huxley model. The vital thing is that the FitzHugh–Nagumo model is simple enough to be analysed mathematically but complicated enough to exhibit some (but not all) of the interesting behaviour of the Hodgkin–Huxley model.

The Hodgkin–Huxley model is a system of four highly non-linear differential equations. However, the essential feature of this system is the cellular membrane potential (6.3). The effect of the Eqns. (6.4), (6.5) and (6.6) is to modulate the membrane potential in a nonlinear fashion and at different relative time scales: some faster than others (depending on the particular ion specie). The FitzHugh–Nagumo equations are a system of two first-order differential equations with only one non-linear term:

$$\frac{dv}{dt} = f(v) - w + I_a, \tag{6.7}$$

$$\frac{dw}{dt} = bv - \gamma w, \tag{6.8}$$

$$f(v) = v(a - v)(v - 1), \tag{6.9}$$

where $0 < a < 1$, $0 < b$ and $0 < \gamma$. The variable v is analogous to the cellular membrane potential, and the variable w is the internal dynamical state (which in the Hodgkin–Huxley model are determined by the various ion currents). Hence, the equations for m, n and h have all been replaced by the single variable w in the FitzHugh–Nagumo model. The nonlinearity in Eqn. (6.9) is a polynomial such that $f(v) \to \infty$ for $v \to \infty$, so a very crude (and poor) approximation would be to replace this with an affine increasing line. Of course, the polynomial kink is what provides all the interesting behaviour. In the next section we introduce the tools necessary to determine the location and stability of equations such as these.

TABLE 6.1:
Stability of a fixed
point of a 1-D flow.

$f'(x_0) < 0$	$f'(x_0) > 0$
Stable	Unstable

6.3 Fixed Points and Stability of a One-Dimensional Differential Equation

In Chapter 2 we described various mathematical models for population. Those models were mathematical maps. Here we are dealing with differential equations. Let us quickly consider the situation in one dimension. Then we will return to the FitzHugh–Nagumo system in two dimensions. Consider a system

$$x_{t+1} = f(x_t). \tag{6.10}$$

As we saw in Chapter 2 we can find the fixed point(s) x_0 by solving $f(x) = x$. The stability of the fixed point x_0 is determined by the value of the linearisation of the system at the fixed point: i.e. if $|f'(x_0)| < 1$ then the fixed point is stable; if $|f'(x_0)| > 1$ then it is unstable (see Table 6.1).

Now, for differential equations, the system is similar but subtly different. Consider

$$\frac{dx}{dt} = f(x). \tag{6.11}$$

The fixed points x_0 are those values of x that don't move; in a system like Eqn. (6.11) this means that $\frac{dx}{dt} = 0$. That is, the fixed points x_0 are solutions of $f(x) = 0$. Just as with maps, to understand whether a given fixed point is stable or unstable we need to look at the linearisation about that fixed point,

$$f(x) = f(x_0) + f'(x_0)(x - x_0) + \frac{1}{2!}f''(x_0)(x - x_0)^2 \tag{6.12}$$

$$+ \frac{1}{3!}f'''(x_0)(x - x_0)^3 + \dots.$$

Note that Eqn. 6.12 is of exactly the same form as Eqn. (2.6). If $x \approx x_0$, then $(x_t - x_0)$ will be small and (provided the higher-order derivatives are bounded), higher-order terms will vanish. The dominant term becomes the first derivative and we can neglect the rest (compare to Eqn. (2.7)),

$$\frac{dx}{dt} \approx f'(x_0)(x - x_0). \tag{6.13}$$

Now, whether the point x approaches x_0 or not depends on $f'(x_0)$. If $f'(x_0) < 0$ and $x > x_0$, then the right-hand side will be negative and x will get smaller

— that is, it will move toward x_0. Conversely, if $f'(x_0) < 0$ and $x < x_0$, then the right-hand side will be positive and x will move up — toward x_0. Because this is a differential equation, the system is a continuous flow. If it continues to move upward, it will eventually reach the fixed point (and vice versa). However, if $f'(x_0) > 0$, then the reasoning is reversed and x will move away from x_0. Table 6.1 summarises this result — and to a large extent this table is simpler than Table 2.1. Since we are dealing with a one-dimensional flow, it must be continuous.

Of course, in two and more dimensions, the situation becomes more complex. To deal with the added complexity of increasing dimensions, we will introduce a graphical analysis scheme. In two dimensions (at least) it is possible to get a fairly good understanding of the dynamical behaviour of such systems from studying the nullclines in a two-dimensional phase-plane.

6.4 Nullclines and Phase-Planes

We will take the two dimensional FitzHugh–Nagumo equations (Eqns. (6.7), (6.8), (6.9)) as a specific example (and the case in which we are ultimately interested). However, for clarity we begin with the general system of two ordinary differential equations:

$$\frac{dx}{dt} = g(x, y) \tag{6.14}$$

$$\frac{dy}{dt} = h(x, y). \tag{6.15}$$

Of course, Eqns (6.7) and (6.8) are a special case. The two-dimensional *phase plane* is the set of points (x, y), such that for each point, one can associate a velocity $(\frac{dx}{dt}, \frac{dy}{dt})$. That is, at every point in this two-dimensional space (x_1, x_y) one can compute the corresponding derivative — the velocity vector $(g(x_1, y_1), h(x_1, y_1))$.

Fixed points of this two-dimensional system may be determined, by solving a system of two equations, in just the same way as we described for one-dimensional equations (Section 6.3). That is, a fixed point (x_0, y_0) of Eqns. (6.14) and (6.15) will satisfy

$$0 = g(x_0, y_0)$$
$$0 = h(x_0, y_0).$$

The two lines $g(x, y) = 0$ and $h(x, y) = 0$ in the phase-plane are called *nullclines* because points on the x-nullcline $g(x, y) = 0$ are only moving in the y direction (since $\frac{dx}{dt} = 0$). Similarly, points on the y-nullcline $h(x, y) = 0$ are only moving in the x direction. The first step toward understanding a system

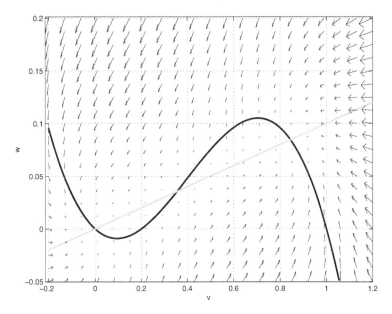

Figure 6.4: Nullclines and phase-plane for the FitzHugh−Nagumo system. The line (grey) and cubic (heavy black) illustrate the w and v nullclines of the FitzHugh–Nagumo equations with $a = 0.2$, $b = 0.05$, $\gamma = 0.5$ and $I_a = 0$. Note that the three fixed points occur at the intersection of these two curves. Numerical computation of $\left(\frac{dw}{dt}, \frac{dv}{dt}\right)$ are marked at grid points over phase space (shown as small arrows). This illustrates the direction of flow at certain points — toward the left- and right-most fixed points, with decreasing velocity.

such as Eqns. (6.14) and (6.15) is to plot the nullclines in two dimensions. This gives us sufficient information to find the fixed point and also begin to understand the stability of points in the neighbourhood of those fixed points.

Figure 6.4 illustrates the phase-plane and nullclines for one particular instance of the FitzHugh–Nagumo system. The nullclines (shown in Fig. 6.4) are the solution of $\frac{dv}{dt} = 0$ and $\frac{dw}{dt} = 0$. First, the v nullcline:

$$\frac{dv}{dt} = 0$$
$$f(v) - w + I_a = v(a - v)(v - 1) - w + I_a$$
$$= 0$$
$$w = v(a - v)(v - 1) + I_a. \tag{6.16}$$

Second, the w nullcline:

$$\frac{dw}{dt} = 0$$
$$bv - \gamma w = 0$$
$$w = \frac{b}{\gamma}v. \tag{6.17}$$

Combining Eqns. (6.16) and (6.17) we find the fixed points of the system satisfy

$$v(a - v)(v - 1)\frac{b}{\gamma}v + I_a = 0. \tag{6.18}$$

Hence, the system will have 1 or 3 fixed points, depending on the solution of the cubic equation (6.18). Of course, it is also possible to study the stability of these fixed points (just as we did in Chapter 2 for simpler maps), but this turns out to be a little complicated for the time being (we will pursue this approach for certain population systems in Chapter 7). Certainly, computer algebra packages could be employed — but even so, the full solutions (for arbitrary a, b, γ and I_a) stretched over 16 pages[2].

Nonetheless, by studying the interaction of the two nullclines, and by computational analysis of the local vector field (that is, the dynamics of sample trajectories near the fixed points), we can learn a lot about the behaviour of the system. In Fig. 6.5 we redraw Fig. 6.4 with sample trajectories as well.

Clearly, the behaviour of the FitzHugh–Nagumo system depends on the manner of intersection of the two nullclines, and the stability of the resultant fixed points. By altering I_a it is possible to raise or lower the cubic curve relative to the line. This applied current I_a can therefore be used to control the number of fixed points that the system exhibits and therefore its dynamic behaviour. In the next section we consider this in more detail.

6.5   Pitchfork and Hopf Bifurcations in Two Dimensions

In Chapter 2, the term *bifurcation* was used to describe the way in which the dynamical behaviour of a system can change in response to the change of a parameter. In that case, the system was described by the logistic map; it had a single parameter (r) and the resulting bifurcation is known as a *period doubling bifurcation*. Now we will study the bifurcation behaviour of the FitzHugh–Nagumo equations in response to changes in the applied current

---

[2]Based on the Mathematica computation for the stability of the fixed points from Eqn. (6.18).

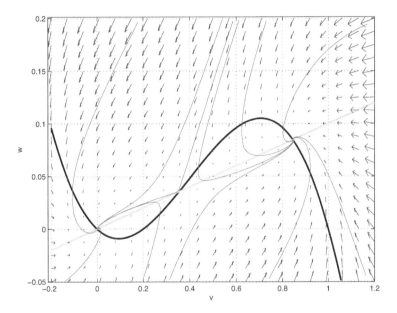

Figure 6.5: **Nullclines, phase-plane and trajectories for the FitzHugh−Nagumo system**. Nullclines and phase-plane are illustrated identically to Fig. 6.4. In addition, we have computed sample trajectories for initial conditions in the vicinity of the fixed points. Note that these trajectories always converge to one of the two exterior fixed points (one trajectory can be observed very nearly reaching the middle fixed point before being deflected away). From this it is clear that the left- and right-most fixed points are stable, and the central one is unstable. Moreover, trajectories crossing the v nullcline are always vertical (since $\frac{dv}{dt} = 0$), and trajectories will always cross the w nullcline horizontally.

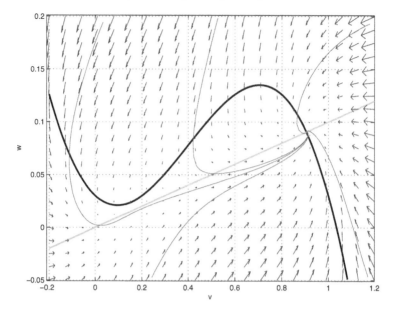

Figure 6.6: Nullclines, phase-plane and trajectories for the FitzHugh−Nagumo system ($I_a = 0.03$). We repeat the illustration of Fig. 6.5, now with $I_a = 0.03$. For these parameter values, the system exhibits only one fixed point — a stable fixed point.

I_a. This current can be thought of as the input to the cell we are modelling; as the magnitude of that input changes, the cell can exhibit one of several different behaviours.

Let us consider the FitzHugh–Nagumo equations with the parameter values of the previous section ($a = 0.2$, $b = 0.05$, $\gamma = 0.5$). By changing I_a we can alter the system such that the cubic and the line only intersect once — see Fig. 6.6. In this case, we go from a single stable fixed point, to two stable and one unstable point. Pictorially we can draw the change in the system behaviour as a function of r as shown in Fig. 6.7. Because of the resemblance of this diagram to a fork, bifurcations of this type are referred to as *Pitchfork bifurcations*.

One important feature of this system is that as the system evolves the stable fixed points never occur "next" to one another (and similarly for the unstable ones). Roughly speaking, if we draw a closed curve around a region of phase space such that the vector field on the boundary of that curve is everywhere pointing inward, then if the interior of that region only has isolated fixed points, the number of stable fixed points will exceed the number of unstable fixed points by one — this is an application of the more general Poincaré–Hopf Theorem.

For the arrangement of cubic and line in Fig. 6.6 (and its predecessors),

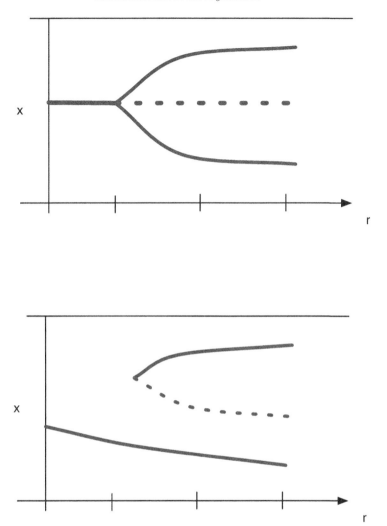

Figure 6.7: **The Pitchfork bifurcation.** Stable fixed points are illustrated as a solid line and unstable ones with a dotted line. Notice that as the bifurcation parameter r increases (in the FitzHugh–Nagumo equations, $r = I_a$), the system goes from one stable to two stable and one unstable fixed points. In the upper panel, we illustrate the classical caricature of a Pitchfork bifurcation (it looks like a pitchfork). The lower panel is what actually happens for the FitzHugh–Nagumo system at these parameter values: as the cubic "touches" the line and then passes through it, it gives rise to a new pair of fixed points, one stable and one unstable. The one-dimensional representation (in terms of a fictitious variable x) of the two-dimensional (w, v) phase space is just to add visualisation: the true situation is too complex to draw in only two dimensions.

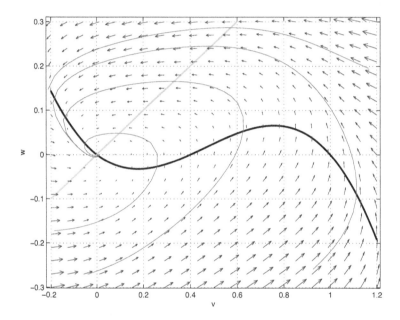

Figure 6.8: **Nullclines, phase-plane and trajectories for the FitzHugh−Nagumo system** ($a = 0.4$, $b = 0.1$, $\gamma = 0.2$ and $I_a = 0$). We repeat the illustration of Fig. 6.4 with diffferent parameter values — such that the slope of the line is steeper than the parabola. For these parameter values, the system will never exhibit three fixed points. The effect of increasing I_a is shown in Fig. 6.9.

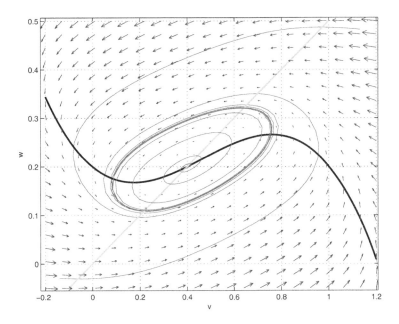

Figure 6.9: **Nullclines, phase-plane and trajectories for the FitzHugh−Nagumo system ($I_a = 0.2$)**. By increasing the applied current I_a, the situation in Fig. 6.8 is modified such that the fixed point at the original becomes unstable and it is now surrounded by a stable limit cycle: point from either outside or inside converge to this stable oscillation.

one can see from purely geometric arguments that the system can go from one to three fixed points (and then back again). However, by changing the slope of either the line or the cubic, it is possible to obtain an arrangement (see Fig. 6.8) where the system will always (for any value of applied current I_a) only have one fixed point. Nonetheless, an interesting bifurcation occurs in this case too. As the value of I_a is increased, the fixed point changes from stable (Fig. 6.8) to unstable (Fig. 6.9) — roughly when the fixed point occurs in the middle section of the cubic. For this range of values of I_a, the fixed point is now encircled by a stable periodic orbit. Initial conditions outside that periodic orbit converge down to it. Initial conditions inside the periodic orbit (and near the fixed point) expand and converge to the periodic orbit. Such trajectories are illustrated in Fig. 6.9.

The conditions necessary for either Hopf or pitchfork bifurcation can be better understood by building one more model. Figure 6.10 is a piecewise affine[3] model of the FitzHugh–Nagumo model (of the Hodkin-Huxley model of the squid giant axon model...).

While this range of dynamics which we have observed is mathematically neat, it also is physiologically relevant: by changing the applied current, one can generate a spontaneously oscillating cell (the Hopf bifurcation) or give rise to stable non-zero fixed points (Pitchfork bifurcation). Both types of behaviour are to be expected in neuronal dynamics. In addition, neurons also exhibit another dynamical behaviour which can be characterised by the simple FitzHugh–Nagumo model: excitability.

6.6   Excitability

One of the most important features of neural activity is *excitability*. When stimulated at a level lower than a certain threshold, the neuron will behave in a quiet state and generate a small *sub-threshold* response. However, for stimulation above a certain threshold, the neural response will be a large excursion in phase space. In Figs. 6.11 and 6.12 we can see this behaviour in the FitzHugh–Nagumo model. In Figs. 6.11 the trajectory (the time traces for $w(t)$ and $v(t)$ are also shown) exhibits a very large excursion (in terms of both magnitude and time) before returning to the stable equilibrium. A slightly smaller excitation (Fig. 6.12) illicits only a very small response.

The excitability observed in Figs. 6.11 and 6.12 is a model for generation and propagation of an action potential (see Section 5.2). Suppose a neural cell exists in a state something like that depicted in Fig. 6.11. Naturally, the cell will converge to the stationary rest state: i.e. the stable fixed point.

---

[3]That is, made up only of straight lines.

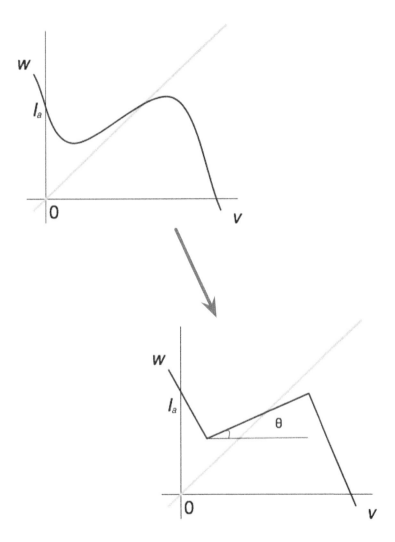

Figure 6.10: **Local affine approximation to the FitzHugh−Nagumo system.** The w and v nullclines of the original system are illustrated in the upper panel. We approximate this with the local affine model in the lower panel and find that three fixed points (and hence a Pitchfork bifurcation) can only occur if $\tan\theta > \frac{b}{\gamma}$ — that is, if the slope of the centre affine section is steeper than the slope of the linear nullcline.

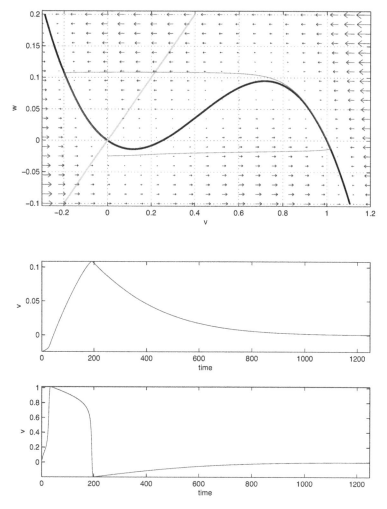

Figure 6.11: **Nullclines, phase-plane and trajectories for the FitzHugh−Nagumo system ($a = 0.25$, $b = 0.001$, $\gamma = 0.002$ and $I_a = 0$).** We repeat the illustration of Fig. 6.4 with diffferent parameter values — such that the system exhibits excitability. For one set of initial conditions (shown here), the system displays a very large deviation in w and v (time traces of $w(t)$ and $v(t)$ are shown as the lower two panels) and a very long slow transient prior to convergence to the fixed point. For a very slightly different initial condition, see Fig. 6.12, the system exhibits a small deviation in w and v and a relatively quick return to the fixed point.

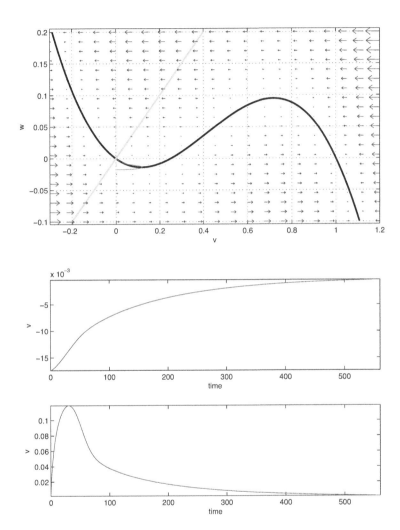

Figure 6.12: Nullclines, phase-plane and trajectories for the FitzHugh−Nagumo system ($a = 0.25$, $b = 0.001$, $\gamma = 0.002$ and $I_a = 0$). Compare with Fig. 6.11. By very slightly changing the initial conditions, the system settles (relatively) quickly to the stable fixed point without the large excursion evident in Fig. 6.11.

Incoming signals to the cell will cause a perturbation to that rest state (as in both Fig. 6.11 and Fig. 6.12). Depending on the magnitude of that incoming signal, the state of the cell may be perturbed further from its rest state (Fig. 6.11) or less far (Fig. 6.12). Then, provided that perturbation is large enough, the cell generates an action potential: that is, the large deviation evident in Fig. 6.11. Moreover, for sub-threshold excitation, no such action potential is observed (Fig. 6.12). The nature of the system also means that a much larger excitation, while generating a slightly large action potential, will not generate an excitation substantially different from any other excitation which exceed the cellular excitation threshold.

In other words, these differential equations can generate a discrete reaction: either action potential occurs or it does not. Of course, for a continuous system of equations, there is actually no such thing as a sharp (discontinuous) boundary. Nonetheless, the boundary is such that very small variation in excitation will lead to one of two very similar reactions. Finally, observe that by altering the applied current I_a (that is, by affecting the cell's background electrochemical environment), it is possible to either enhance or suppress this entire behaviour.

6.7   Summary

The Hodgkin–Huxley differential equations provide a physiological model of cellular dynamics. The FitzHugh–Nagumo equations are an abstraction of the higher-dimensional Hodgkin–Huxley system. While the variables in the FitzHugh–Nagumo model are now abstracted from any specific physiological meaning (roughly v is something like membrane potential and w is something like all the internal variables rolled into one), the dynamics that this system displays are still physiologically meaningful: spontaneous oscillation, stability and excitation.

To understand these dynamics we have employed a variety of approximation and analytical techniques, together with computation exploration. We could find fixed points, and observe that it would be possible — albeit complicated — to analytically compute their stability. In the next chapter, we look at a different biological system: one with simpler dynamics which can more easily be studied analytically.

Glossary

Axon A cable-like extension of a nerve cell (or neuron) which provides a mechanism by which the electrical signal of the cell can be transmitted from the cell body. In particular, the early experiments of the squid giant axon chose this particular nerve cell as it offered a large and accessible cell structure which could be analysed and manipulated experimentally.

Conductance The propensity of a medium to conduct electricity. By definition this is the reciprocal of the electrical resistance.

Excitability A dynamical system which, when perturbed over a certain level, exhibits a large activity before returning to the stable state. The certain level is called the threshold.

FitzHugh–Nagumo equations A computationally and almost analytically tractable simplification of the Hodgkin–Huxley equations which, nonetheless, retains many of the most interesting dynamically behaviours of the more complicated equations.

Hodgkin–Huxley equations Based on their early work on squid giant axons, the Hodgkin–Huxley equations provide a mathematical description of the chemical and (hence) electrical dynamics within nerve cells.

Hopf bifurcation A bifurcation under which a single fixed point spawns a limit cycle (which initial has radius zero), surrounding a new fixed point. If the initial single fixed point is stable, then so is the limit cycle, while the new fixed point (interior to the limit cycle) is unstable. Or, vice versa.

Nullcline A curve in the phase-plane representing the set of points (that is, systems states) such that the rate of change of the system in a particular direction is zero.

Phase-plane A geometrical-mathematical representation of the state of a system. If a system can be described by a set of n variables (for example, via an n-dimensional system of differential equations), then the n-dimensional representation of these variables constitute the phase-plane. As the single set of these variables (corresponding to a particular realisation of the system) change in time, this will result in a curve through the n-dimensional space: a trajectory.

Piecewise affine A function made up of a series of straight line segments.

Pitchfork bifurcation A bifurcation under which by changing a parameter value the system goes from exhibiting one fixed point to three. This can either be super-critical (one stable fixed point becomes two stable and one unstable fixed points) or sub-critical (on unstable fixed point becomes two unstable and one stable fixed points).

Sub-threshold Excitation below the threshold is called sub-threshold.

Exercises

1. Obtain and read Hodgkin and Huxley's original papers (there were three of them).

2. Why are the conductance terms said to be of the form $(A + Be^{-Ct})^D$?

3. Plot $(A + Be^{-Ct})^D$ and explain the general shape in terms of the parameters A, B, C and D.

4. Assuming steady-state voltage (not necessarily a good assumption), find a solution for n (in $\frac{dn}{dt} = \alpha_n(1 - n) - \beta_n n$).

5. Therefore, explain the possible shape (assuming steady-state voltage) for n^4 and $m^3 h$.

6. Consider the FitzHugh–Nagumo (FHN) system

$$\frac{dv}{dt} = v(a - v)(v - 1) - w + I_a$$

$$\frac{dw}{dt} = bv - \gamma w$$

with $a = 2$, $b = 0.8$ and $\gamma = 5$.

   (a) At these parameter values, what sort of bifurcation will occur as I_a changes.

   (b) Determine the number of fixed points the system has for $I_a = 0$

   Provide analytic justification for your answers.

7. (a) Consider the FHN equations with $a = 0.2$, $b = 0.0025$ and $\gamma = 0.005$. With no external current, describe the dynamical behaviour of the system.

   (b) Plot typical time traces for this system for different initial condiions. Use these time series to illustrate that the dynamics are indeed as claimed in part (a). (HINT: You'll need to modify my source code, available from the website, or invoke one of the MATLAB® ode-solvers — you could try ode45.)

    (c) For the same initial conditions, describe the effect of introducing an applied current of 0.03.

    (d) What effect does reversing this current have?

8. TheFHN model of cellular dynamics is given by

$$\frac{dv}{dt} = f(v) - W + I_a$$

$$\frac{dw}{dt} = bv - \gamma w$$

$$f(v) = v(a - v)(v - 1)$$

Suppose for a particular cell, $a = 0.2$, $b = 0.02$ and $\gamma = 0.1$.

    (a) If the magnitude of the applied current is 0, describe the dynamics.

    (b) If the applied current is 0.03 (dimensionless units), what are the dynamics now?

    (c) Keeping the applied current constant, what change needs to be made to change the dynamics between Hopf-type and Pitchfork-type bifurcations?

9. Formulate the piecewise linear approximation for the FHN equations, and restate the condition $\tan\theta < \frac{b}{\gamma}$ in terms of the original FHN nullclines.

10. What is the required condition for three fixed points in the FHN system? (HINT: You'll need to use the formulae for solution of a cubic.)

11. Formulate the conditions for bifurcation (the existence of three fixed points) in terms of the original FHN equations. (HINT: See the previous exercise.)

12. Use the program **fhn_field.m** (available from the website) to compute the vector field near the bifurcation point for both Pitchfork and Hopf type bifurcations.

13. Use the program **fhn_field.m** with system parameters $a = 0.25$, $b = 0.002$, $\gamma = 0.002$, $I_a = 0.0$ to find the separatrix of excitability. That is, find two initial conditions as close together as possible which exhibit sub-threshold and super-threshold responses. Describe the origin of this separatrix.

14.   (a) Use the MATLAB® program **fhn_field.m** to determine the location (parameters) of the onset of the Pitchfork bifurcation of the FitzHugh–Nagumo system described in the text. At some point the systems goes from one, to two and then to three fixed points. What appears (from these simulations) to be the stability of each fixed point in each case?

(b) Provide analytic support of your conclusions in Exercise 14a.

15. Consider the Bonhoeffer–van der Pol oscillator given by the following equations

$$\frac{dx}{dt} = x - \frac{x^3}{3} - y$$

$$\frac{dy}{dt} = (x - a) - y$$

(a) For $a = 1$, compute and then sketch the nullclines.

(b) Hence, with reference to the phase diagram in the previous part question, describe qualitatively how the dynamics will change as a changes.

(c) What neuro-physiological phenomenon does this behaviour carica-ture?

(d) Now consider a neural model described by the following two differ-ential equations,

$$\frac{dx}{dt} = f(x, y, I_a)$$

$$\frac{dy}{dt} = g(x, y, I_a),$$

where $f(x, y, I_a) = g(x, y, I_a) = 0$ at only one point (x_0, y_0). Let

$$A(x, y) = \begin{bmatrix} \frac{\partial f}{\partial x} & \frac{\partial f}{\partial y} \\ \frac{\partial g}{\partial x} & \frac{\partial g}{\partial y} \end{bmatrix}.$$

If the eigenvalues of $A(x_0, y_0)$ are $\lambda_1 = I_a$ and $\lambda_2 = I_a + 1$, de-scribe the nature of the bifurcations of this system in terms of the parameter I_a.

Chapter 7

Population Dynamics

7.1 Predator–Prey Interactions

In Chapter 2 we met the logistic (difference) equation and explored the population dynamics of a single specie. In this chapter we extend these ideas. We start with two competing species: a predator specie P (conventionally, "foxes") and a prey specie N (usually, "rabbits").

In the absence of foxes, the rabbits will breed and their population will increase (just as it did for Fibonacci) at a rate proportional to their number. Foxes, however, will eat the rabbits at a rate proportional to both their populations. Conversely, in the absence of food (rabbit), the fox population will decrease. Once the foxes have something to eat, their population increases at a rate proportional to the product of the two populations. Hence, we have

$$\frac{dN}{dt} = N(a - bP) \tag{7.1}$$

$$\frac{dP}{dt} = P(cN - d), \tag{7.2}$$

where a, b, c and $d \geq 0$ are all positive constants. Note that if $b = 0$, then the fox population will be subject to Malthusian population decay; and if $c = 0$, the the rabbit population will grow exponentially (see Section 2.1). In fact, there is a close connection between these equations and those of Malthus and Verhulst. The predator-prey equations (7.1 and 7.2) were proposed by Vito Volterra in 1926 to study fish population dynamics. The equations of Volterra are themselves based on a model of chemical reactions proposed by Alfred Lotka. Lotka's equations, in turn, are a variation of the logistic equation.

Referring to Eqns. (7.1) and (7.2), with the four constants a, b, c and d all non-negative, one can see that each of the four terms on the right-hand side of the equations has a specific meaning. The aN term is the natural growth of the prey population, and the $-dP$ term is the natural decay (famine) of the predator in the absence of prey. The interaction terms $-bNP$ and cNP model the benefit of the presence of prey to the predators, and the harm to the prey caused by the predators[1]. Finally, we note that by changing the signs

---

[1] Only under certain very specific situations would we expect that $b = c$.

of the constants, it is possible to model various other ecological interactions.

For example, suppose that N and P represent two species in competition for the same resources. Then in the absence of that competition, either specie would flourish (i.e. $a > 0$ and $d < 0$), but the presence of each specie is detrimental to the other ($b > 0$ and $b < 0$). Similarly, symbiosis[2] can also be modelled. Each specie will grow by itself (except in the most extreme form of interdependence) and so $a > 0$ and $d < 0$. However, each specie also benefits from the presence of the other: $b < 0$ and $c > 0$.

Nonetheless, let us consider the general system of Eqns. (7.1) and (7.2) which, if $a, b, c, d > 0$, models predation. The natural question is whether, and if so when, will both species exist in a happy equilibrium.

7.2 Fixed Points and Stability of Two-Dimensional Differential Equations

The fixed points of Eqns. (7.1) and (7.2) can be obtained as follows (just as we did in Section 6.4):

$$\frac{dN}{dt} = 0$$
$$= N(a - bP)$$
$$\frac{dP}{dt} = = 0$$
$$= P(cN - d)$$

and hence

$$aN = bNP \tag{7.3}$$
$$dP = cNP. \tag{7.4}$$

Hence, from Eqn. (7.3), $N = 0$ or $P = \frac{a}{b}$ and from Eqn. (7.4), $P = 0$ or $N = \frac{d}{c}$. Therefore, the system has two fixed points (if $b > 0$ and $c > 0$):

$$(N_0, P_0) = (0, 0) \quad \text{and} \quad \left(\frac{d}{c}, \frac{a}{b}\right).$$

Now, to analyse the stability of these fixed points, it is possible to follow exactly the procedure in Section 6.3 –but one must replace the one-dimensional representation with the n-dimensional equivalent. If x is the n-dimensional

---

[2]Two species in a mutually beneficial arrangement. For example, bees and flowers.

state variable of a system of differential equations,

$$\frac{dx}{dt} = f(x),\tag{7.5}$$

with $f : \mathbf{R}^n \to \mathbf{R}^n$ and x_0 is a fixed point (hence $f(x_0) = 0$), then we may apply the n-dimensional version of the Taylor series expansion (see Eqn. 2.13):

$$\frac{dx}{dt} = f(x_0) + (x_t - x_0)^T \nabla f(x_0) + \dots,$$
$$= (x_t - x_0)^T \nabla f(x_0) + \dots,\tag{7.6}$$

where $(\cdot)^T$ is the vector transpose operator and $\nabla f(x)$ is the gradient of f (see Eqn. 2.14). As in Chapter 2, we will truncate the expansion at the first derivative. Then we let $A = \nabla f(x_0)$ and assume that the matrix A has full rank and that it is therefore diagonalisable. Hence $A = PDP^{-1}$, where P is invertible and D is a diagonal matrix. Specifically,

$$P = \begin{bmatrix} v_1 \ v_2 \ \cdots \ v_n \end{bmatrix}$$

$$D = \begin{bmatrix} \lambda_1 & 0 & \cdots & 0 \\ 0 & \lambda_2 & \cdots & 0 \\ \vdots & \vdots & \ddots & \vdots \\ 0 & 0 & \cdots & \lambda_n \end{bmatrix},$$

where $v_1, v_2, \dots, v_n$ are the eigenvectors of A and $\lambda_1, \lambda_2, \dots, \lambda_n$ are the eigenvalues.

Now, Eqn. (7.6) can be re-written as

$$\frac{dx}{dt} = (x - x_0)PDP^{-1}$$
$$\frac{d(xP)}{dt} = (xP - x_0P)D.\tag{7.7}$$

Hence, we can perform the change of variables $\xi = xP$ and we have

$$\frac{d\xi}{dt} = (\xi_t - \xi_0)D,\tag{7.8}$$

where, since the matrix D is diagonal, the original d-dimensional map (7.6) is now a system of d one-dimensional maps (7.8). When some of the eigenvalues are complex, this is not entirely possible; one will get pairs of equations which cannot be decoupled (as real numbers). Each of these equations is of the form

$$\frac{d\xi^{(i)}}{dt} = \lambda_i \xi^{(i)}(t) - \xi_0^{(i)} \lambda_i,\tag{7.9}$$

where $\xi^{(i)}$ is the i-th component of ξ (i.e. $\xi = (\xi^{(1)}, \xi^{(2)}, \dots, \xi^{(n)})$). The solution of this equation (7.9) is simply $\xi^{(i)}(t) = Ae^{\lambda_i t} + \xi_0^{(i)}$ – which converges to $\xi_0^{(i)}$ if $\lambda_i < 0$ (i.e. it is stable) and is unstable if $\lambda_i > 0$.

Hence the stability of the system (7.5) can be determined by finding the fixed points x_0 (which satisfy $f(x_0) = 0$), and computing the eigenvalues λ_i and eigenvectors v_i of the matrix $A = \nabla f(x_0)$. If $\lambda_i < 0$, the system is stable in the v_i direction, and if $\lambda_i > 0$, it is unstable.

We can now return to Eqns. (7.1) and (7.2) as a specific example. In this case,

$$f(x) = \begin{bmatrix} aN - bNP \\ cNP - dP \end{bmatrix} \tag{7.10}$$

$$\nabla f(x) = \begin{bmatrix} a - bP & -bN \\ cP & cN - d \end{bmatrix}. \tag{7.11}$$

Let us consider the fixed point $(N, P) = (0, 0)$ in this case,

$$\nabla f(x_0) = A$$
$$= \begin{bmatrix} a & 0 \\ 0 & -d \end{bmatrix}.$$

The next step is to take the matrix A and compute its eigenvectors and eigenvalues so that it can be diagonalised. Of course, in this case it is trivial (since A is already diagonal). We have

$$\lambda_1 = a$$
$$\lambda_2 = -d$$
$$v_1 = \begin{bmatrix} 1 \\ 0 \end{bmatrix}$$
$$v_2 = \begin{bmatrix} 0 \\ 1 \end{bmatrix}.$$

Hence (if $a, d > 0$, as they would be for the case of predation), the origin $(0, 0)$ has one stable ($[0, 1]^T$) and one unstable ($[1, 0]^T$) direction – that is, it is a saddle point. In two dimensions, the types of fixed points one may observe mirror those for a two-dimensional difference equation. Hence Fig. 7.1 mimics Figs. 2.13 and 2.14.

Now for the second fixed point $\left(\frac{d}{c}, \frac{a}{b}\right)$ of Eqns. (7.1) and (7.2). Substituting $(N_0, P_0) = \left(\frac{d}{c}, \frac{a}{b}\right)$ into Eqn. (7.11) gives

$$\nabla f(x_0) = A$$
$$= \begin{bmatrix} 0 & -\frac{bd}{c} \\ \frac{ac}{b} & 0 \end{bmatrix}.$$

The eigenvalues of A can be determined as the solution of

$$\det(A - \lambda I) = 0$$
$$\begin{vmatrix} -\lambda & -\frac{bd}{c} \\ \frac{ac}{b} & -\lambda \end{vmatrix} = \lambda^2 + ad$$
$$\lambda^2 = -ad.$$

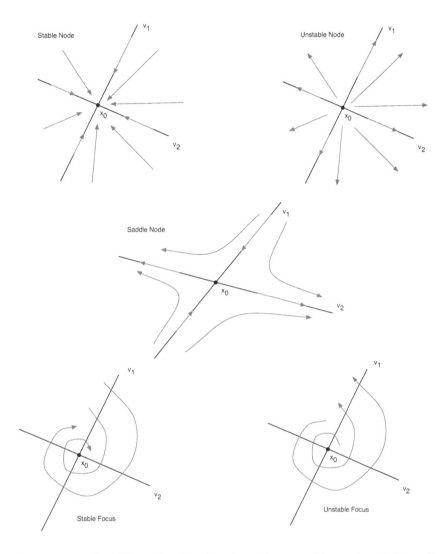

Figure 7.1: **Stability of a fixed point of a two-dimensional flow.** In two dimensions, the fixed point x_0 will have two eigenvalues λ_1 and λ_2 with associated eigenvectors v_1 and v_2. If $\lambda_{1,2} < 0$, the fixed point is a stable node. If $\lambda_{1,2} > 0$, the fixed point is an unstable node. If $\lambda_1 < 0 < \lambda_2$ (or vice versa), then the fixed point is a saddle, with a stable direction v_1 and an unstable direction v_2. If λ_1 and λ_2 form a complex conjugate pair, then the system will exhibit a focus. If the real part of $\lambda_{1,2}$ satisfies $\text{Re}(\lambda_{1,2}) < 0$, the fixed point is a stable focus. Otherwise, if $\text{Re}(\lambda_{1,2}) > 0$, the fixed point is an unstable focus.

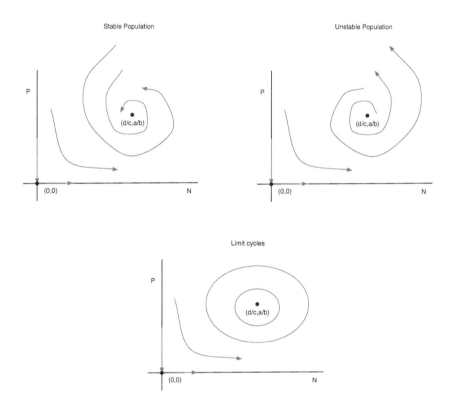

Figure 7.2: **Stability of the predator-prey system Eqns. (7.1 and 7.2).** The system has two fixed points at $(0,0)$ and $(\frac{d}{c}, \frac{a}{b})$. In the absence of prey, the predators will die out; and in the absence of predators, population of prey will explode – hence the origin is a saddle point. If the population of predators is too large, it will decrease towards zero and then the population of prey will start to increase. The stability of the second fixed point cannot be resolved as it has purely imaginary eigenvalues. The two conventional possibilities are a stable focus (if the real part was negative) or an unstable focus (if it is positive). In the stable case, the population would oscillate and then converge to the equilibrium value. In the unstable case, the population would grow without bound — leading to very large predator populations followed by a crash and an explosion in prey and so on. Finally, it is also possible that there exist stable (or unstable) limit cycles: that both predator and prey exist in a balanced but oscillating equilibrium with each population periodically waxing and waning.

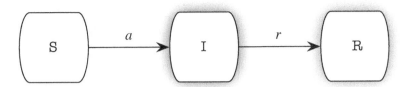

Figure 7.3: **SIR disease model.** Individuals may be susceptible (S), infected (I) or removed (R), and transitions between these states have rates dictated by the parameters a and r.

Since $a, d > 0$ (again, assuming predation), we have eigenvalues which form a complex conjugate pair $\lambda_{1,2} = \pm i\sqrt{ad}$. If the real part is negative, then they form a stable focus, if not, then the focus is unstable. Unfortunately, in this case, the eigenvalues are purely imaginary – they have zero real part. Therefore no conclusion can be drawn. Rather, we are left with one of three possibilities – a stable focus, an unstable focus, or limit cycle solutions. In Fig. 7.2 we illustrate these possibilities (we do not have sufficient evidence to determine which is true).

7.3 Disease Models: SIS, SIR and SEIR

We now turn to one specific type of predator prey system. In this case, the predator is an infectious agent and the prey is the host. Typically, one models this system by considering a number of host individuals. They may be susceptible to a disease (S), infected with a disease (I) or removed from the population (R). Individuals may arrive at the third R state because they have recovered from the disease and are now immune, they have been isolated (quarantined) from the rest of the population, or they died (presumably from the disease). Susceptible individuals become infected with some rate r and infected individuals are removed at a rate a. Hence, we have the following equations:

$$\frac{dS}{dt} = -rSI \tag{7.12}$$

$$\frac{dI}{dt} = rSI - aI \tag{7.13}$$

$$\frac{dR}{dt} = aI. \tag{7.14}$$

The rate r is referred to as the *infection rate* and a is the *removal rate*. The model was proposed in 1927 by A.G. McKendrick and W.O. Kermack to explain epidemic diseases such as plague and cholera. It is therefore widely referred to as the Kermack–McKendrick model as well as the SIR model; see Fig. 7.3.

Notice that the equations contain a symmetry such that $\frac{dS}{dt} + \frac{dI}{dt} + \frac{dR}{dt} = 0$ and hence the total population $N = S + I + R$ is constant. Suppose that the initial conditions are $S(0) = S_0 > 0$, $I(0) = I_0 > 0$ and $R(0) = 0$ (anyone that has already been removed from the population is of no consequence to the model). The most basic thing we can wish to learn from this model is, under what situation will the disease spread?

First, notice that $\frac{dS}{dt} \leq 0$ for all t (that is, there can never be an increase in the susceptible population). Hence, $S(t) \leq S_0$ for all $t > 0$. Now $\frac{dI}{dt}\big|_t = 0 = I_0(rS_0 - a)$ and this expression will be larger than 0 only if $S_0 > \frac{a}{r}$. Moreover, if $S_0 < \frac{a}{r}$, the $\frac{dI}{dt} = I(rS - a) < 0$ and the infected population will decrease and continue to decrease. Hence, an epidemic (defined here as an initial increase in the number of infected individuals) will occur only if $S_0 > \frac{a}{r}$. It is common when referring to these equations to define the *relative removal rate*,

$$\rho = \frac{a}{r}, \tag{7.15}$$

and the *reproduction rate*,

$$R_0 = \frac{rS_0}{a}, \tag{7.16}$$

where now, if $R_0 > 1$ an epidemic will occur.

The SIR model of Kermack–McKendrick is intended to model disease outbreaks in situations where either the disease is fatal (and hence R refers to death), or where recovery from the disease confers immunity (one can only get infected once). In situations where infection can occur repeatedly, we need to use an alternative model, the SI model.

In the SI model, individuals can be only susceptible or infected. Susceptible individuals become infected at some rate r and infected individuals become susceptible at a rate a (Fig. 7.4):

$$\frac{dS}{dt} = -rSI + aI \tag{7.17}$$

$$\frac{dI}{dt} = rSI - aI. \tag{7.18}$$

The term rSI reflects the rate of infection (which is proportional to the size of both susceptible and infected populations) and aI is the rate at which infected individuals recover. As before, the total population $N = S + I$ is constant,

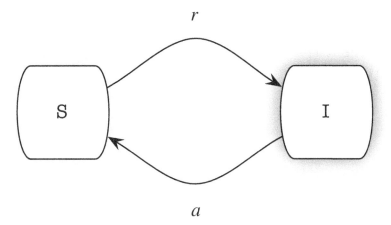

Figure 7.4: **SI (or SIS) disease model.** Individuals may be susceptible (S), infected (I), and transition between these two states have rates dictated by the parameters a and r.

and hence Eqns. (7.17) and (7.18) can be rewritten as a single equation:

$$\frac{dI}{dt} = r(N - I)I - aI$$
$$= -rI^2 + (rN - a)I. \tag{7.19}$$

Note that since $-rI^2 < 0$ (for all t) if $(rN - a) < 0$, then $I \to 0$. Otherwise Eqn. (7.19) has two steady states: when $\frac{dI}{dt} = 0$ either $I = 0$ or $I = N - \frac{a}{r}$. Let $f(I) = -rI^2 + (rN - a)I$. Since $f'(I) = -2rI + (rN - a)$ and

$$f'(0) = rN - a$$
$$f'(N - \frac{a}{r}) = -rN + a,$$

we find that if $\frac{a}{r} > N$, then the steady state at the origin is stable and the one at $N - \frac{a}{r}$ is unstable (this is what we can observe directly by looking at the sign of the right-hand side of Eqn. (7.19)). Conversely, if $N > \frac{a}{r}$, then the steady state at the origin is unstable and the one at $N - \frac{a}{r}$ is stable.

Hence the dynamics of the SI equations (7.17 and 7.18) can be summarised as follows. For $\frac{a}{r} > N$ there is a stable equilibrium at zero (and a negative unstable equilibrium which does not concern us) and the epidemic will die out. For $\frac{a}{r} < N$ the steady state at the origin becomes unstable and there is a stable equilibrium at $I = N - \frac{a}{r}$. Hence if the population N is sufficiently large (for a given a and r), then the disease will persist in a proportion of the population. It is interesting, and perhaps surprising, that the success of an

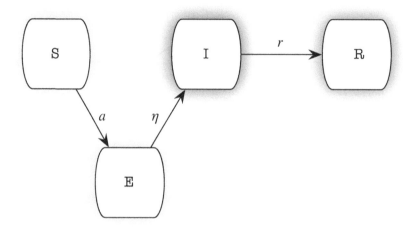

Figure 7.5: SEIR disease model. Individuals may be susceptible (S), exposed (E), infected (I) or removed (R). Transition between these states have rates dictated by the parameters a, η and r. The distinction between the SIR and SEIR models is the inclusion of a latent state E with some transition rate η.

epidemic depends on the size of a population (since a and r are properties of the disease and independent of the population). The critical threshold $N > \frac{a}{r}$ indicates that for well-mixed populations of a certain size, the disease will continue. Hence, eradication of a disease of this form is possible by reducing the effective population size N — that is, by reducing the degree to which people mix.

The SIR model is appropriate for disease for which infection confers immunity (or results in certain death). For disease where re-infection is possible, the SI (sometimes called the SIS) model is more appropriate. However, for some diseases, there is a latency period during which time an individual is infected, but not infectious. Examples of this type of disease would be HIV/AIDS and SARS. In both cases, upon infection, an individual will carry the disease for some time (days in the case of SARS, but years in the case of AIDS) before becoming symptomatic and infectious. Let E denote the population of individuals which is infected, but not yet infectious. We can extend the SIR model (the SIS can also be extended in a similar way) to a system of four equations:

$$\frac{dS}{dt} = -rSI \tag{7.20}$$

$$\frac{dE}{dt} = rSI - \eta E \tag{7.21}$$

$$\frac{dI}{dt} = \eta E - aI \tag{7.22}$$

$$\frac{dR}{dt} = aI. \tag{7.23}$$

The new term ηE is the rate at which infected individuals become infectious and therefore the parameter η determines the latency period for a given disease (Fig. 7.5). As mentioned, it turns out that this is a good model for transmission of both SARS and HIV/AIDS. We consider both in the following sections.

7.4   SARS in Hong Kong

The Severe Acute Respiratory Syndrome (SARS) virus first appeared during October 2002 in the Guangdong province of southern China. It then passed over the border to Hong Kong and from there spread to Europe, Africa, Asia, Australia and the Americas [4]. The outbreak in 2003 infected 8422, killing 916 [49]. Here we will focus on modelling the transmission of SARS within Hong Kong. Besides mainland China, Hong Kong suffered the greatest casualties [49]. In addition, the epidemiological data available for Hong Kong is certainly more reliable than that of the Chinese mainland[3].

Referring to the data in Fig. 7.6, we will attempt to model this with standard SEIR-type epidemic dynamics (Eqns. 7.20 through 7.23). In the case of transmission of SARS-CoV, $N \gg E + I + R$ and so $N \approx S$. The equations then reduce to

$$\frac{dE}{dt} = aNI - \eta E$$
$$\frac{dI}{dt} = \eta E - rI \tag{7.24}$$
$$\frac{dR}{dt} = rI.$$

---

[3]During the epidemic, mainland authorities classified information on SARS infections as a state secret. Moreover, bureaucracy caused much of the available information to be concealed. Despite this, subsequent official investigation indicates that the infection rate for the Chinese mainland during the later part of the outbreak was apparently significantly over-reported.

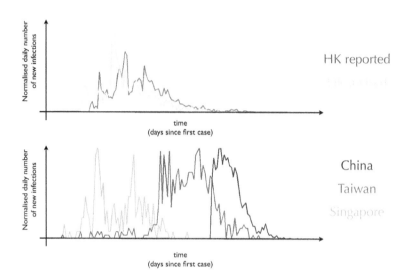

Figure 7.6: **Daily reported SARS case in Hong Kong (HK) and Asia.** At the end of the epidemic, 105 days after the first recorded case, a total of 1755 confirmed infections had been identified in Hong Kong. The reported data are the daily number of hospital admission of SARS cases, as published in the *South China Morning Post* (and other local media) during the outbreak (starting on February 15 2003). Subsequent to the epidemic (several months later), a revised set of data was released by the local government health authority. This is the "revised" data also shown in the upper panel (with a more pronounced spike). The lower panel represents data from Taiwan, Mainland China (i.e. excluding Hong Kong, Macau and also Taiwan) and Singapore. The data for China is rather unreliable (characterised as it is by a step change — corresponding to the last peak in the lower panel), and outbreaks in both Singapore (the earliest outbreak on the lower panel) and Taiwan were relatively limited in scope. See colour insert.

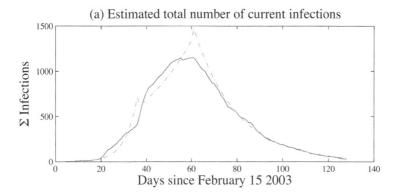

Figure 7.7: Estimate of the total daily number of infectious individuals for HK. The solid line is the cumulative sum of the revised daily SARS case data minus the cumulative sum of the daily recoveries and fatalities. The dashed line is three exponential fits to this data as described in the text. (Figure reproduced from [38].)

Since the available data are only at the resolution of individual days, it is perhaps more useful to re-write this as a set of equivalent difference equations,

$$\Delta E_t = aN_tI_t - \eta E_t$$
$$\Delta I_t = \eta E_t - rI_t \tag{7.25}$$
$$\Delta R_t = rI_t,$$

which can be easily solved. Solutions are either exponential growth or exponential decay. From the time series data for transmission of SARS-CoV in Fig. 7.6, we must derive a value for I. The data available in Fig. 7.6 is *reported* cases; this is neither equivalent to new infection nor new exposures (it is actually closer to new removals, since individuals when reported are isolated – but, as it turns out, quarantine was not perfect so this is not true either).

We approximate the number of infectious individuals as the cumulative sum of the (revised) number of SARS cases minus the cumulative sums of the number of deaths and recoveries. This time series is depicted in Fig. 7.7. In making this approximation we assume that hospitalised individuals are still infectious (this assumption is unavoidable, we will see later that it is also reasonable), and that recovered individuals are no longer infectious.

The disease now follows essentially two phases: a growth phase and a decay phase. Clearly from about day 60 the number of infections is decreasing. It is easy to fit an exponential decay curve to these data and therefore obtain a value for $\rho = \frac{a}{r}$. During the growth phase there are actually two distinct sub-phases. During the initial spread of the disease (up to about day 37), spread is uncontrolled and rapid. Between day 38 and 60, spread continues, but at a slowing rate. It therefore makes sense to fit distinct values of ρ to these two sections of data. From Fig. 7.7, we show the exponential fit to these three phases. Of course, the fit could be improved by having an increasing number of phases – or even making $\rho(t)$ a function of time (and changing every day). However, this is undesirable because then the model becomes more complicated than the data it is trying to describe! In Chapter 10 we will present an alternative solution.

The problem with using a model such as the SEIR model is that it assumes that all individuals are mixing equally and that anyone in the community could be infected. Particularly for disease where only a relatively small portion of the population is actually infectious, this may not be a good assumption. An alternative is to build a network to model the contacts between individuals and to then model the disease spread on the network. Surprisingly, it turns out that this can be done in a very economical way (that is, it is a very simple model) and it provides a very good model of data such as this (see Section 10.2)).

7.5   Summary

Diseases have long been successfully modelled as discrete compartments. Individuals can either be susceptible to a disease, infected with that disease or recovered, for example. Using these compartments, it is relatively straightforward to write down a series of difference or differential equations based on the rate at which individuals transit between these states. Having done this, modelling disease transmission is not significantly different from any model of population growth — such as those in Chapter 2. In particular, in Section 2.4 we introduced a model of hormone secretion and chemical control of cancer. That model is basically the same as those introduced here — with the added complication that it exhibited a piece-wise switch between two states (medication being either on or off).

Nonetheless, with such a simple model we observed almost "unreasonably"[4] good results. This very simple model is able to accurately model spread and growth of a wide range of phenomena, including disease spread. However, we see that this model is not always sufficient. For a disease, spreading among a relatively small sub-population, the assumption of homogeneous mixing (that is, the assumption that all individuals are equally likely to be infected) breaks down. Then it becomes much more important to track which individuals are

---

[4]The term "unreasonable" is borrowed from Cheryl Praeger, who in turn, borrowed it from Eugene Wigner.

actually infected. To deal with this situation it becomes necessary to model the actual connections between individuals — this is a topic which we will defer until Chapter 10. Meanwhile, in Chapter 8 we will show how the simple models of this chapter can be extended to model the spatial spread of disease.

Glossary

Competition When two species compete for a common food source. The growth in population of each specie is detrimental to the other.

Infection rate The rate at which susceptible individuals within a community are infected with a disease. The corresponding (within a fixed interval of time) probability of an infection.

Kermack–McKendrick model The mathematical equations describing a SIR disease transmission model.

Lotka–Volterra equations The particular set of differential equations describing predator-prey interactions.

Predation The archetypal predator-prey system. One specie feeds of the other. The growth of the prey specie population leads to the growth of the predator population which, in turn, leads to decline in the prey specie and corresponding decline in the predator population.

Predator-prey systems A system examining the interaction between predator and prey species.

Relative removal rate The ratio of removal rate to infection rate.

Removal rate The rate at which infected individuals within a community are relieved of that infection (either cured or killed). The corresponding (within a fixed interval of time) probability becoming no longer infected.

Reproduction rate The ratio of the initial susceptible population to the relative removal rate.

SEIR model A SIR disease model with the addition of an exposed state. Such models are useful for modelling diseases where infection does not immediately lead one to being infectious. Hence, infected individuals first enter the exposed state, and only once they become infectious do they move to the "I" state.

SI model A particular disease model where individual can only be susceptible or infected (there is no recovery).

SIR model A particular disease transmission model where individuals can be susceptible "S", infectious "I" or removed "R". Transition is between these three groups in that order. Susceptible individuals become infected and then removed: hence for a given individual infection can only occur once.

SIS model A particular disease model where individuals transit between infected and susceptible states. Infected individuals, when recovered, become susceptible once again.

Symbiosis When two species co-exist in a mutually beneficial system. The growth in population of each specie is beneficial to the other.

Well-mixed population A well-mixed population is one for which all individuals are assumed to be equally likely to come into contact with all others. Hence, there is a uniform rate of transmission across the entire population.

Exercises

1. In the standard SIR model given above, what proportion of a population needs to be effectively vaccinated to remove an endemic infection?

2. It is possible to re-write the SIR differential equations as a series of three difference equations:

$$S(t+1) = S(t) - \hat{r}S(t)I(t)$$
$$I(t+1) = I(t) + \hat{r}S(t)I(t) - \hat{a}I(t)$$
$$R(t+1) = R(t) + \hat{a}I(t).$$

   Relate $\hat{r}$ and $\hat{a}$ to the original co-ordinates. Show that the total population will be constant. Now it is possible to solve this difference equation by re-writing it in matrix form.

3. For the original SIR differential equations, what is the significance of the removal rate $\rho = \frac{a}{r}$ and the reproduction rate $R_0 = \frac{rS_0}{a}$? Explain why it is that a disease which would lead to an epidemic in a large population fails to do so in a smaller one? Isn't this somewhat anti-intuitive? Finally, show that if $rS < a$, then the extent of an outbreak is decreasing (being controlled).

4. An outbreak in a population of 6.7×10^6 has an infection rate $r = 10^{-7}$ and a removal rate $a = 0.1$. Does this outbreak qualify as an epidemic? How can a be modified to control the outbreak?

5. Consider a three-compartment (S, I and R) model of disease transmission. Let the rate of infection be 1×10^{-4} and the rate of removal 0.1.

   (a) Construct (i.e. state) the differential equations corresponding to this model.

   (b) For what values of S_0 does this disease formally constitute an epidemic?

   (c) Is the infection growing or shrinking?

   (d) Solve these differential equations (with $N = 10000$, $S_0 = 20$ and $R_0 = 0$) using MATLAB® .

   (e) For this population, what value of rate of removal provides the boundary between a disease that is growing without bound (endemic) and one which will eventually terminate?

   (f) Re-write these differential equations as equivalent inter-day difference equations for the same model.

6. Consider two species interacting as predator and prey (i.e. "predation", for example rabbits R and foxes F). The population over time of these two species is given by

$$\frac{dR}{dt} = \alpha R - \beta RF$$

$$\frac{dF}{dt} = \gamma RF - \delta F,$$

where the growth rate for rabbits is $\alpha > 0$ and the natural decay rate for foxes is $\delta > 0$.

   (a) Find the fixed points of this system. For what values of the parameters are the two fixed points (call them (R_0, F_0) and (R_1, F_1)) physically relevant? Describe the meaning (in terms of rabbits and foxes) of each fixed point.

   (b) Linearise the right-hand side of the differential equations about the two fixed points $(R_{0,1}, F_{0,1})$ to obtain expressions of the form

$$\begin{bmatrix} \frac{dR}{dt} \\ \frac{dF}{dt} \end{bmatrix} = \begin{bmatrix} R_{0,1} \\ F_{0,1} \end{bmatrix} + A \begin{bmatrix} R_{0,1} - R \\ F_{0,1} - F \end{bmatrix} + (\text{higher} - \text{order terms}).$$

The "higher-order terms" correspond to the non-linear part and the matrix A (which you must determine) is the derivative of the system equations with respect to the two variables R and F evaluated at the fixed point.

(c) Determine the stability of the two fixed points of the linearised system. Remember, for example, that the fixed point (R_1, F_1) will be stable if the real parts of the eigenvalues of the matrix $A|_{(R,F)=(R_1,F_1)}$ are less than zero. That is, if the real part of the eigenvalue is positive, then that eigendirection is unstable, and vice versa.

(d) Computationally verify the stability of the fixed points for appropriate values of the system parameters. Explain any deviation between theory and computation.

7. For an infectious agent with a relative removal rate of 10^4, what is the minimum population required to support an epidemic? Explain the significance of this number.

8. The Hong Kong flu H3N2 outbreak of 1968 and 1969 killed an estimated one million people worldwide. However, the effect of the disease was mitigated by the fact that previous exposure to other N2 flu strains (such as the Asian flu on the 1950s) resulted in a higher tolerance to infection with the Hong Kong strain. Infected individuals become infectious immediately and spread the virus through the air via coughing and sneezing. Moreover, once infected, recovery would occur in 4 to 5 days, with such exposure conferring complete immunity. Clinical data show that the death rate among infected individuals over the age of 65 was significantly higher than the general population. Transmission of Hong Kong flu can be modelled with the following system of equations,

$$\frac{dS_1}{dt} = -r_1 S_1 I$$

$$\frac{dS_1}{dt} = -r_2 S_2 I$$

$$\frac{dI}{dt} = r_1 S_1 I + r_2 S_2 I - aI$$

$$\frac{dR}{dt} = aI,$$

where S_1 is the number of susceptible individuals without previous exposure to N2 flu, S_2 is the number of susceptible individuals with previous exposure to (another) N2 flu strain, I is the number of infected individuals and R is the number of removed individuals. The rates of infection are $r_1 > r_2 > 0$ and the rate of removal is a. Finally, suppose that the two susceptible populations are in an initial ratio $h = \frac{S_1(0)}{S_2(0)}$.

(a) Why is an SIR model (rather than either SIS or SEIR) appropriate in this case?

(b) What effect does increased mortality among those aged over 65 have on the model, and why?

(c) Derive a condition for the disease becoming endemic in terms of only the system parameters and initial conditions stated above.

(d) Using your answer from Exercise 8(c), introduce an expression for the total susceptible population $S(t) = S_1(t) + S_2(t)$ and show that changing h can prevent an epidemic. Hence, state the necessary condition on h to prevent an outbreak.

(e) In a population of 50000, 30% have been exposed to a previous N2 strain. If $r_1 = 10^{-5}$ and $r_2 = 10^{-7}$, how large does a have to be to prevent an outbreak?

Chapter 8

Action, Reaction and Diffusion

8.1 Black Death and Spatial Disease Transmission

The disease models introduced in Chapter 7 assume that the population is "well-mixed". That is, all individuals in the community interact in such a way that any infected individual is equally likely to infect every other member of the community. For very many diseases this is plainly not the case. In Section 7.4 we saw that such models could only give an approximate description of transmission of SARS in Hong Kong, and only if the infection rate was allowed to vary. However, since SARS only affected a small proportion of the population, the actual contact between infected individuals and the general community was limited to a particular sub-community. In Section 10.2 we will introduce a better model which accounts for actual contact between individuals in the wider community. In this section we consider another special type of disease transmission process. In this case, individuals are subject to some infectious agent, but the individuals are arranged spatially — that is, they can only infect others in the same, or neighbouring geographical region.

This model is particularly useful for modelling the spatial spread of diseases over large regions, and, in particular the spread of diseases such as bubonic plague. Figure 8.1 depicts historical spread of bubonic plague through Europe in the fourteenth century. The model is essentially an extension of the ones presented in Chapter 7 for disease transmission. For simplicity we consider the S and I equations of the SIR model and add a term $D\frac{\partial^2}{\partial x^2}$ for spatial diffusion,

$$\frac{\partial S}{\partial t} = -SI + D\nabla^2 S \tag{8.1}$$

$$\frac{\partial I}{\partial t} = SI - \lambda I + D\nabla^2 I \tag{8.2}$$

$$\lambda = \frac{a}{rS_0}, \tag{8.3}$$

where D is the rate of diffusion, and the parameter λ is introduced to simplify the above expression. As we saw in the previous chapter, the SIR model maintains a constant total population N, and hence, here we ignore the R population as it can be deduced directly from Eqns. (8.1) and (8.2). Moreover,

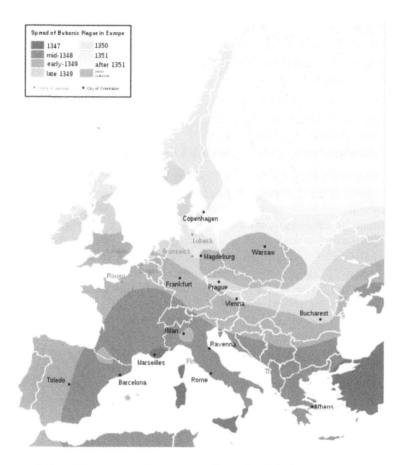

Figure 8.1: Plague in Europe. Spread of bubonic plague in four-teenth century Europe can clearly be seen to propagate as a wave across the continent, originating in a port in northern Italy and spreading throughout continental Europe and Britain. In medieval Europe, bubonic plague was fatal within 2 or 3 days in roughly 80% of cases. Hence, during the period 1347–1351, plague spread across Europe and killed at least one quarter of the entire population (con-servative estimates tend to exceed 20 million). (Image obtained from http://commons.wikimedia.org/wiki/File:Bubonic_plague_map_2.png   and freely distributed under the GNU Free Documentation License (GFDL)).)

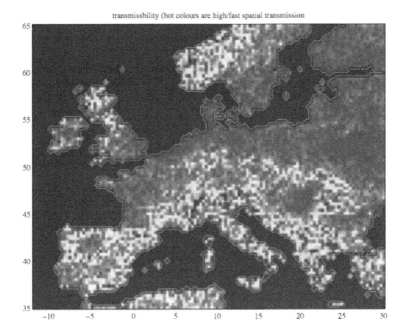

transmissbility (hot colours are high/fast spatial transmission

Figure 8.2: Approximate diffusivity across fourteenth century Europe. Terrain roughness is used as an approximate estimate of diffusivity. The second derivative of altitude is used to estimate the difficulty of travel in a given location. Flat terrain is easier to travel over and rough terrain is more difficult.

in Eqns. (8.1) and (8.2) we assume that there is the same diffusion rate D for both populations S and I. It may be easier to simplify these equations a little further. Let us assume that the spatial diffusion is symmetric — that is, occurs the same in each direction. Then we may think of Eqns. (8.1) and (8.2) as equivalent to a one-dimensional model

$$\frac{\partial S}{\partial t} = -SI + D\frac{\partial^2 S}{\partial x^2} \tag{8.4}$$

$$\frac{\partial I}{\partial t} = SI - \lambda I + D\frac{\partial^2 I}{\partial x^2}, \tag{8.5}$$

where the population variables $S(x,t)$ and $I(x,t)$ diffuse in one spatial direction x.

Although these nonlinear partial differential equations can be somewhat difficult to solve, they are amenable to computational analysis. Suppose, for added realism, we assume that the diffusion parameter D is actually a function of space, and we take the diffusion to be dependent on the terrain. Certainly, in medieval Europe diffusion would have been more rapid on flat terrain and

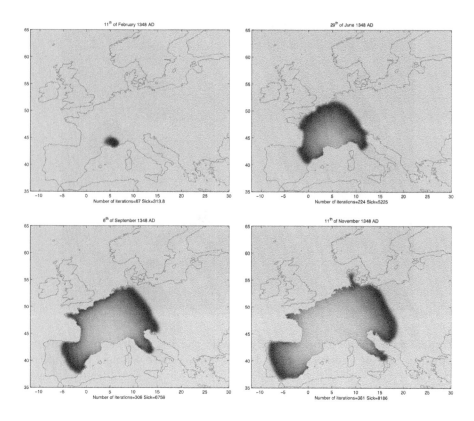

Figure 8.3: **Computational simulation of spreading plague.** Density of infected individuals is shown superimposed on a map of Europe. From an initial infection in Northern Italy, the disease spread in a wave across (modern) France and then through the Iberian peninsula, and central and eastern Europe. Compare this to Fig. 8.1.

more limited in mountainous regions. Then we can obtain a rough estimate of diffusitivity for fourteenth century Europe $D(x, y)$ as in Fig. 8.2.

With this estimate of diffusivitiy $D(x, y)$ it is possible to generate a fairly realistic simulation of the spread of plague in the fourteenth century — as illustrated in Fig. 8.3. Just as for the historical data in Fig. 8.1, the spread first covers the area of modern France and the Iberian peninsula and central and eastern Europe. Unlike the historical data, the computer simulation shows a slightly slower spread through Italy — possibly due to the same bacteria entering through other Italian ports, or more rapid travel (and therefore diffusion) through the Italian peninsula. From Fig. 8.2 our estimate of diffusion rate for the region of modern Italy is relatively slow. However, as this was then part of the unified Holy Roman Empire, this estimate may be rather too

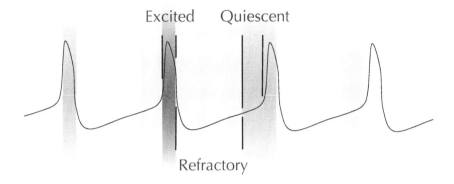

Figure 8.4: **Reaction diffusion.** The individual parts of this spatial system can exist in one of three states — *excited*, *quiescent* or *refractory*. After excitation, the system remains excited for some period and then enters and must remain in the refractory state for some fixed time. Then, in the quiescent state, the system is ready to be excited again.

low. Similarly, we estimated in the model a relatively retarded transmission from continental Europe to Britain — historical data do not support this.

Nonetheless, small modifications show that with a relatively simple diffusion model it is possible to have an accurate model of the spread of plague — this is despite the fact that this model assumes a uniform population density across Europe. For other diseases, such as SARS and avian influenza, despite having an abundance of data, this is not possible. In Section 10.2 and Section 10.3 (Chapter 10), we introduce new models to cope with these situations.

8.2 Reaction–Diffusion

In this section, we extend the general diffusion model of Section 8.1 to a more generic environment. In Section 8.1 the model of disease transmission works as a spatial model because of the diffusion term. For many different physical systems, this diffusion is accompanied by a *refractory* (or resting) period and such systems are known as *reaction-diffusion* — see Fig. 8.4. Such systems can either be *excited* (if they have been activated), *refractory* (if they are resting, following an excitation), or *quiescent*. The excited state persists for a short time and must be followed by a fixed (and longer) time in the refractory state. Following the refractory period the system is quiescent until it is again activated and enters the excited state.

Systems such as this are very easy to simulate computationally. To do so,

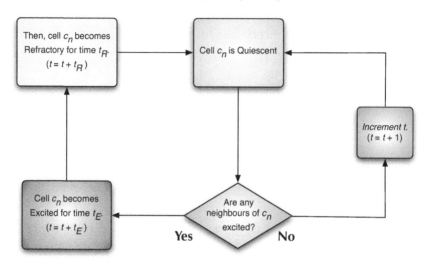

Figure 8.5: **Reaction diffusion.** The individual parts of this spatial system can exist in one of three states — *excited*, *quiescent* or *refractory*. Transition between these states for a single cell are depicted graphically.

we can construct an array of cells; each cell is connected to its neighbours and occupies one of these three states. Excitability is propagated by an excitable cell to any neighbours which are quiescent (and therefore ready to be excited). Such simulations can be extremely instructive and can be made to mimic a wide range of phenomena. The update rule for each cell can be summarised with the following pseudo-code:

1. Suppose that at time t cell c_n is quiescent. Symbolically, $c_n(t) = Q$ (each cell can be in one of three states E, R or Q).

2. If any of the neighbours of c_n, c_{NN}, are excitable, then $c_n(t+1) = E$.

3. Increment t.

4. If $c_n(t) = E$, then $c_n(t+\delta) = E$ for $0 < \delta < t_E$ and then $c_n(t+\delta) = R$ for $t_E < \delta < t_R + t_E$ (where t_E is the duration of the excitable phase and t_R is the duration of the refractory phase). Then let $t = t + t_E + t_R$ and $c_n(t) = Q$.

5. Now the cell c_n is currently quiescent. Return to Step 1.

The same sequence is depicted in Fig. 8.5.

Alternatively, we can formulate an explicit partial differential equation model — and in fact, this is precisely the FitzHugh–Nagumo model of Section

6.2:

$$\frac{\partial u}{\partial t} = f(u) - v + D\frac{\partial^2 u}{\partial x^2} \tag{8.6}$$

$$\frac{\partial v}{\partial t} = bu - \gamma v \tag{8.7}$$

$$f(u) = u(a - u)(u - 1), \tag{8.8}$$

where u is the excitability variable (for example, cellular membrane potential), v captures the internal dynamics, D is the spatial diffusion constant, and $a < 1$, b, and γ are all positive. Comparing this to Eqns. (6.7), (6.8) and (6.9), the only distinction is the replacement of the external activation current I_a with the diffusion term. Here, the individual cells receive diffusive activation from their neighbours.

Models such as this can be employed to model a wide range of phenomena — including, in particular, the excitation of cardiac tissue and the heart beats, behaviour of crowds (as individuals react to their neighbours) and even the spread of a forest fire or flash flooding.

8.3 Cardiac Dynamics

Given the model in the previous section it is now straightforward for us to produce computational simulations for various cardiac dynamics. Let us consider a rectangular grid of cells, where each cell has four neighbours (horizontally and vertically, but not diagonally) for simplicity. In addition to the cellular automata reaction-diffusion model of Fig. 8.5, we also introduce one additional rule: suppose that propagation of excitability is not perfect. That is, if the neighbour of a quiescent cell is excitable, then the quiescent cell becomes excitable with some probability p (which is close to 1). Then, if p is *close enough* to 1, we see regular propagation of excitation as depicted in Fig. 8.6.

The simulation in Fig. 8.6 shows waves of excitation (and then relaxation passing over the two-dimensional medium at regular intervals — exactly analogous to the regular excitation and propagation of excitatory waves in cardiac tissue). Moreover, just like with cardiac tissue, we can also explore what happens when the simulation breaks down. In Fig. 8.7 we show the same simulation at four later times. In this figure we see the onset of a high-frequency oscillation due to the development of a secondary pacemaker centre. This secondary centre is actually a group of ordinary cells which supports a stable wave propagating around itself as a stable spiral. As the spiral oscillates, it continually interacts with newly quiescent cells, generating a continuing wave of excitation which then propagates out as concentric tightly packed rings of excitation. Eventually (in Fig. 8.8), this regular excitatory spiral breaks down

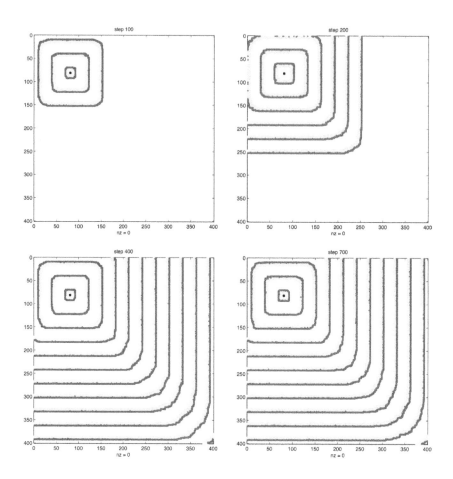

Figure 8.6: Cellular automata simulation of regular cardiac dynamics. A group of cells in the upper right quadrant (shown as a small black dot) is made to pulse periodically (every 30 time steps), generating an excitatory wave of duration 2 time steps followed by a refractory period of 6 time steps. This simulation results in regular plane waves propagating from the pacemaker cells (periodically pulsing) over the full extent of the simulation. Because of the (artificial) square geometry (with neighbours vertically and horizontally), the simulation has an artificial square-ish shape. The fact that the corner of this simulated wavefront appears rounded is that we impose an imperfect propagation rule — that is, a neighbour of an excitatory node becomes excited with some probability p less than 1 (in this simulation $p = 0.75$). Figures 8.7 and 8.8 are a continuation of this simulation under the same conditions.

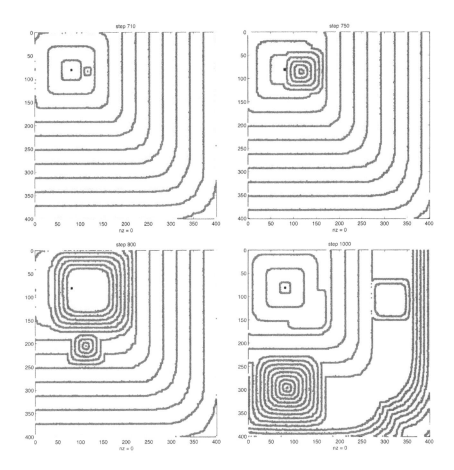

Figure 8.7: **Cellular automata simulation of regular cardiac dynamics, showing periodic "tachycardic" breakdown.** Starting in the top right plot, a small number of cells remain quiescent while an excitatory waves passes and then becomes excited in the tail of that wave. This creates a backward propagating wave (i.e. moving toward the pacemaker cells). At the core of this wave is a small spiral which rotates around itself and continually enters newly quiescent cells and generates waves of excitation. Eventually (in Fig. 8.8), this spiral fails to continue to propagate and the high-frequency excitatory waves are again replaced by the dominant low-frequency regular oscillation from the pacemaker cells.

Figure 8.8: **Cellular automata simulation of regular cardiac dynamics — recovery of sinus rhythm.** In this continuation of the previous plot, we see the waves of high-frequency excitation grow and completely dominate the entire simulation medium. Then, these waves eventually fail to be self-sustaining, die down and are replaced by the resumption of the regular wave..

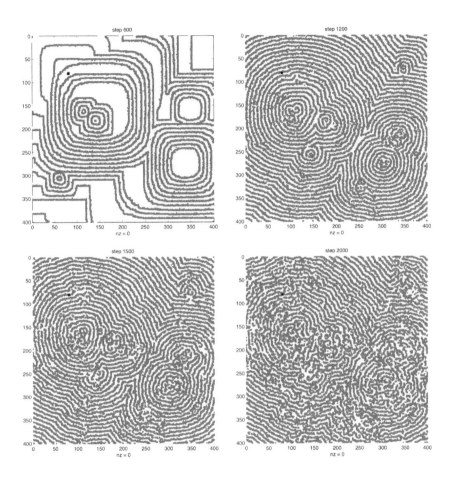

Figure 8.9: **Cellular automata simulation of fibrillatory cardiac dynamics.** By lowering the probability of successful propagation slightly (from $p = 0.75$ in the previous simulation to $p = 0.68$ here), we see that high-frequency oscillations persist and eventually fracture and spread throughout the medium, leading to a fully developed state of spatial turbulence (the final plot). The simulation, and how it is generated, are analogous to the onset of ventricular fibrillation and cardiac instability.

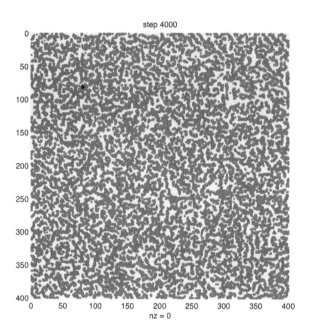

Figure 8.10: **Cellular automata simulation of fibrillatory cardiac dynamics — fully developed turbulence.** Continuing the simulation of the previous figure, the system eventually develops into a state of sustained disorder. Waves of excitation propagate and fragment in all directions.

and the original pacemaker-driven waves again dominate. The simulation is exactly analogous to the onset and then self-termination of high-frequency tachycardia in the heart. In the next figure we see what happens when such rhythms do not self-terminate.

In Fig. 8.9 the individual self-sustaining spirals continue for longer and give rise to further spirals. These then interact with one another and lead to a more complicated breakdown in the orderly wave transmission of the previous examples (Figs. 8.6, 8.7 and 8.8). If this situation is allowed to continue, the system eventually evolves into a state of sustained (i.e. continual) spatio-temporal turbulence. The entire media consists of multiple small and disjoint waves moving in various directions. This has a striking similarity to the activity of a fibrillatory heart — something which is described by heart surgeons as feeling like a bag of wriggling worms. Of course, while the heart tissue is in this state, the heart itself is not capable of functioning as a coherent pump. No work gets done and blood does not get transported around the body. Consequently, muscle tissue becomes deprived of oxygen and starts to degrade. Of course, this includes the muscle tissue of the heart itself, which also needs to be supplied with oxygenated blood. Hence, while in the simulation, this state can be sustained indefinitely; in the real heart it cannot.

8.4 Summary

In this chapter we began by extending the disease models of Chapter 7 to spatial transmission (and in particular the example of medieval transmission of bubonic plague) and then showed how the same equations could be used to model a range of systems, including cardiac tissue. The equations we used to achieve this, Eqns. (8.6) and (8.7), are nothing more than the same FitzHugh–Nagumo equations we introduced in Section 6.2 to model neurodynamics. Of course, these equations are non-linear, and particularly when incorporating spatial partial derivatives, can be rather difficult to solve. To circumvent this we introduced a simple computational model and showed how this could model the range of spatio-temporal behaviours evident both in the equations and also the underlying biological system. In the next two chapters we turn our attention to some emerging areas of interest when studying the dynamics of biological systems. In Chapter 9 we explore computational models of large systems of interacting and moving individuals (such as schools of fish or flocks of birds). In Chapter 10 we finish by exploring the problem of modelling the interactions themselves.

Glossary

Cellular automata A computational system simulation consisting of an interconnected array (possibly a grid) of cells. Each cell can occupy one of a finite set of states and that state is updated, in response to the state of itself and its neighbours, at discrete times.

Excitable media A spatial system which can exhibit a large response in reaction to stimulus over some prescribed level, but no response otherwise. Once a response is illicited, there is a prescribed time interval (the refractory period) prior to any further excitation.

Excited The activated state in an excitable medium.

Quiescent The relaxed and ready state in an excitable medium.

Reaction-diffusion The process of reaction (in the sense of an excitable medium) and diffusion (as in spatial spreading).

Refractory The resting, or recovery, state in an excitable medium.

Exercises

1. The FitzHugh–Nagumo equations,

$$\frac{\partial u}{\partial t} = f(u) - v + D\nabla^2 u$$

$$\frac{\partial u}{\partial t} = bu - \gamma v$$

$$f(u) = u(a - u)(u - 1),$$

are used to model reaction-diffusion processes in a wide range of spatial systems (from disease propagation and spread of forest fires, to cardiac instabilities and pattern formation).

(a) Assume that only one spatial direction is relevant and rewrite the equations in an appropriate form.

(b) Briefly describe (in words) why this system provides a good model of both reaction and diffusion in excitable media.

2. Answer the following questions with reference to the cellular automata models of reaction-diffusion. Suppose a given two-dimensional reaction-diffusion system remains excited (when excited) for 6 ms and has a refractory period of 30 ms.

(a) Describe the nature of the system response if it is stimulated with a point source at a frequency of 20 Hz. How long is the (mean) quiescent phase?

(b) Describe the nature of the system response if it is stimulated with a point source at a frequency of 50 Hz. How long is the quiescent phase?

(c) Qualitatively (as accurately as possible, and with the aid of a diagram to illustrate the expected wave propagation) describe the nature of the system response if it is stimulated with two point sources, each with a frequency of 10 Hz and separated by a distance of 10 cm (wave velocity is 0.3 ms^{-1}). Your diagram should, as far as possible, include correct quantitative labels for all features observed.

Chapter 9

Autonomous Agents

9.1 Flocking

In Chapter 8 we looked at various diffusion processes: including several models by which we were able to describe the spatial spread of a disease, or, cellular excitation. In this chapter we look at a different type of spatial interaction: interaction between discrete individuals within a community. Our basic objective is to describe how the individuals within a community interact during movement. There are many different words to describe groups of different types of animals (including people): flock, herd, school, crowd, swarm, and so on. Our objective here is to propose some simple rules for the actions of each individual within the community and show that the collective behaviour of the entire ensemble (flock, herd, school, etc.) behaves in a way consistent with what is observed in nature. For the sake of genericness, here we refer to the individual members of the group as *agents* and the entire ensemble as the *flock* or, occasionally, *collective*. However, the following discussion is not in any way limited to (for example) birds.

When one observes the movement of a school of fish or a flock of birds, the behaviour of the ensemble of animals appears coherent and somehow organised. It is often as if the entire group of animals is behaving collectively as a single entity. An entire school of fish slips around the reef as if it is a single, much larger animal. A flock of birds shifts and darts in the sky as if it is a single giant mass. Yet, in both cases it is clear that movement decisions are being made at the level of individual animals: neither the fish nor the birds really have a single collective brain. Therefore, the primary question is how the movement of individuals within the collective is determined in such a way that the entire ensemble behaves in such a coherent fashion.

But, there is a more subtle question here too. Of course, it is possible to design a set of rules for individual agents such that their collective behaviour appears "naturalistic." One could imagine some relatively straightforward set of rules governing the decision-making process of individual agents (certainly no need to simulate the entire cerebral cortex and central nervous system of each bird — only part of it), together with rules governing aspects of fluid mechanics, aerodynamics and the mechanics of flight (in birds, for example).

Such a system of rules would certainly allow for a sufficient number of parameters so that it could definitely give some realistic simulations, under some circumstances. The real challenge is to determine the minimal set of information which is sufficient. Our aim is to produce the very simplest model of the agent behaviour which is also sufficient to explain the complex collective behaviour observed in nature. As it turns out, a rather rich variety of behaviour can be observed with only the very simplest of all rules. We describe this in the next section.

9.2   Celluloid Penguins and Roosting Starlings

If individual agents are too close, then they should move apart. If they are far apart, then they will move together. Hence, perhaps the simplest rules for autonomous behaviour of agents within a collective (as proposed in [29]) are:

1. Avoid collision: do not get too close to those nearby;

2. Follow: attempt to maintain the same velocity (i.e. both speed and heading) as those around you; and

3. Stay close: try to stay close (but not too close — see Rule 1).

It turns out that implementing these three rules (with the order of precedence given above) actually works surprisingly well. Of course, there are some details in the implementation of these three rules.

All three rules imply some level of local perception on the part of individual agents: either a fixed number of neighbouring agents, or a fixed radius within which neighbours are considered significant. That is, the individual agents do not have a global awareness and will only follow (Rule 2) and approach (Rule 3) certain other flock-mates. Both Rule 1 and Rule 3 require some way to access the position of the rest of the flock. For example, consider three birds in a line: should the middle bird move closer to the left or closer to the right?

First of all, which neighbours do we consider significant? Figure 9.1 illustrates the possible solutions. Of these, the simplest two options[1] consider either all neighbouring agents within a fixed radius, or a fixed number of neighbours. A third alternative is to also model the visual field of each individual agent: it will only determine its position based on those other agents which it can actually "see". This additional complication is useful if, for example, one wishes to model more complex flight formations in groups of birds (such as the typical "vee" formation of a flock of geese). Whatever method one adopts, suppose that the neighbours of a bird i can be listed as $j \in N_i$.

---

[1]There is some debate over which of these two approaches is biologically more relevant.

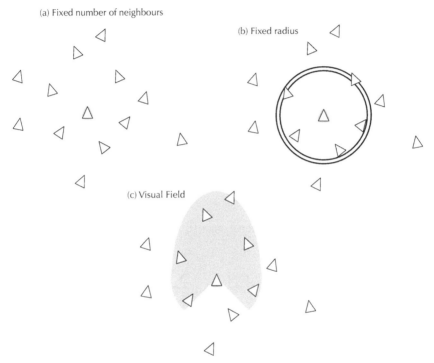

Figure 9.1: Neighbourhood of an agent. Exactly which other agents among the flock should be considered depends on which rule we choose to follow. Here we illustrate: (a) a fixed number of nearest neighbour agents irrespective of actual distance; (b) all agents within some fixed distance (irrespective of number); or (c) all other agents within a fixed "visual field."

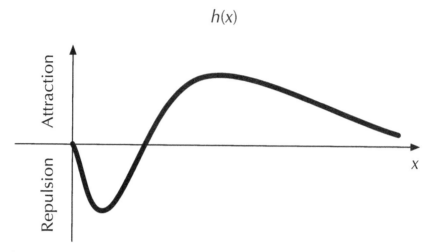

Figure 9.2: **Attraction or repulsion.** Whether to approach one's neighbours or move away from them is determined by how close the agents are. If they are far apart, then the force is attractive; if they become too close, then a repulsive force wins out. A simple analytic expression of this form would be $h(x) = x(x-1)e^{-x}$.

If the position of bird i at time t is $x_i(t)$, then the mean position of the neighbours of bird i is

$$x_{NN(i)}(t) = \frac{1}{|N_i|} \sum_{j \in N_i} x_j,$$

where $|N|$ denotes the number of elements in the set N. Similarly for Rule 2, one needs to evaluate the velocity of the neighbouring agents; the average velocity over all neighbours can be defined in the same way,

$$v_{NN(i)}(t) = \frac{1}{|N_i|} \sum_{j \in N_i} v_j.$$

Now, for each agent $x_i(t)$ we need to calculate a new position $x_i(t+1)$ and velocity $v_i(t+1)$ based on the information available within the collective at time t. The updated position of each agent x_i needs to take account of both its current velocity $v_i(t)$ and also the relative position of the other neighbouring particles $x_{NN(i)}(t)$. The velocity can then also be updated based on the velocity of the neighbours $v_{NN(i)}(t)$,

$$x_i(t+1) = x_i(t) + \tau \left((1-\delta_1)h(\|d_{NN}\|)d_{NN} + \delta_1 v_i(t)\right) \tag{9.1}$$
$$d_{NN} = x_{NN(i)}(t) - x_i(t)$$
$$v_i(t+1) = (1-\delta_2)v_i(t) + \delta_2 v_{NN(i)}(t), \tag{9.2}$$

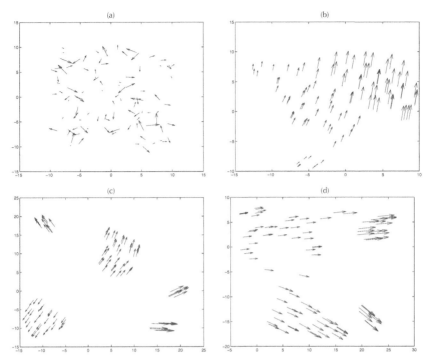

Figure 9.3: **Boid simulations.** The illustrated simulations show three different realisations (b–d) of different random initial conditions (of which panel (a) is typical). In panel (b), the swarm maintains a single unit; and panels (c) and (d) show examples of splitting of the swarm into subunits. The simulations were conducted with 100 agents, $\tau = 0.3$, $\delta_1 = \delta_2 = 0.5$ and by selecting five nearest neighbours for each agent.

where δ_1 and δ_2 are parameters that determine the extent to which the neighbourhood information actually affects the update of both velocity and position. For example, if $\delta_2 = 0$, the velocity remains unchanged and there is no information from the neighbours of an agent being used to determine velocity. Similarly for δ_1. The vector d_{NN} is just notational shorthand for the vector point from the current state $x_i(t)$ to the centre of mass of the neighbours $x_{NN(i)}(t)$. The parameter τ is essentially an integration time step — it controls the speed of update. As $\tau \to 0$, the simulation updates very slowly; if $\tau \gg 0$, the simulation updates much more rapidly. Finally, the scalar function h controls the attraction or repulsion dictated by Rules 1 and 3. If $h > 0$, then agent i will approach the centre of mass of its neighbours (Rule 3); if $h < 0$, then it is repelled (Rule 1). The typical form of h is depicted in Fig. 9.2.

If necessary, one can also add a genuine collision avoidance rule. The implementation described in Eqns. (9.1) and (9.2) does not actually guarantee

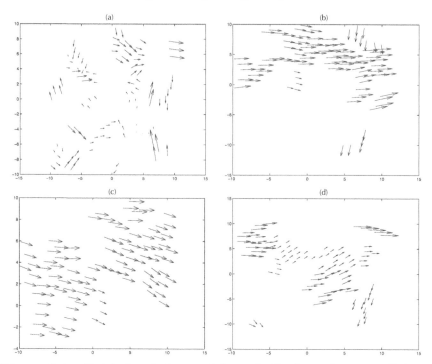

Figure 9.4: Boid simulations with periodic boundary conditions.
The illustrated simulations show four different snapshots at progressively later times of a single simulation. The simulations were conducted with 100 agents, $\tau = 0.2$, $\delta_1 = \delta_2 = 0.1$ and by selecting only one nearest neighbour for each agent. Even with very minimal information exchange, a collective motion is inevitable — because the boundary conditions are periodic and therefore the individual agents eventually all inter-mix with one another.

that agents will not collide, only that they will not get too close to the *average* position of their neighbours. To avoid actual collisions it would also be necessary to check the explicit distance between neighbouring agents: if the distance is less than some minimum threshold, then it should be increased. Figure 9.3 shows flock-like behaviour in some simple simulations according to Eqns. (9.1) and (9.2). By varying the simulation parameters (although, in Fig. 9.3 we did not) one can observe a range of cohesive or individualistic behaviours among the agents.

In Fig. 9.4 we observe that, with one rather biologically dubious assumption, we can virtually guarantee a collective motion — at least if the density of agents is sufficient. We assume that the boundary conditions of the simulations are periodic. This is a fancy way of saying that if agents depart from the box in which we perform the simulation, then they are inserted on the opposite side. That is, we fixed the simulation to run only in a bounded two-dimensional box. Any agent departing from the top boundary is moved to the bottom boundary; similarly, agents leaving on the left reappear on the right, and vice versa. With this assumption we are able to repeat the simulation of Fig. 9.3 with communication only to one neighbour for each agent, and still we obtain a collective coherent motion of the community, as shown in Fig. 9.4. The reason for this is that the density of the agents with the two-dimensional box means that individual groups will inevitably collide and coalesce to form progressively larger groups.

The onset of an "ordered phase" and the transition between order and disorder were also observed in an even simpler modelling of flocking-like behaviour [43]: keep the periodic boundary conditions, but only alter the velocity based on neighbour-neighbour interactions. That is, each agent evaluates the velocity of its neighbours and adopts a velocity similar to the neighbours, with the addition of some noise[2]. Essentially this is a special case of Eqns. (9.1) and (9.2) with $\delta_1 = \delta_2 = 1$:

$$x_i(t+1) = x_i(t) + \tau v_i(t) \tag{9.3}$$
$$v_i(t+1) = v_{NN(i)}(t) + \nu_i(t), \tag{9.4}$$

where $v_{NN(i)}(t)$ is the mean velocity of the neighbours and $\nu_i(t)$ is a noise term. In the original study [43] there is a further simplification: one assumes that $v_i(t) = v$ is constant for all i and t and the update Eqn. (9.4) is actually an update of the velocity heading — that is, the direction:

$$\theta_i(t+1) = \theta_{NN(i)}(t) + \epsilon_i(t), \tag{9.5}$$

where $\theta_i(t)$ is the heading of agent i at time t

$$\theta_i(t) = \arctan\left(\frac{\sin v_i(t)}{\cos v_i(t)}\right)$$

and $\epsilon_i(t)$ is a uniformly distributed noise term.

Having posited suitable — albeit, basic — rules, there are two fundamental questions which we must resolve. The first is "why do this at all?" The second question, which is more difficult to answer, is "how do we know we are right?" The initial impetus to provide models like this was for computer graphics: to provide realistic models of large scale scenes in movies and computer games. In the 1992 movie "Batman Returns"[3], a version of this algorithm was applied to generate realistic crowd scenes. The computer generated crowds of bats and missile-carrying penguins were much cheaper than the massive army of human extras employed 33 years earlier in "Ben-Hur". Fifteen thousand human

---

[2]In fact, in [43] the model is even simpler: all particles are assumed to have the same magnitude of velocity (i.e. same speed) and so it is only the direction of movement of each agent which is updated.

[3]That's the Tim Burton one, featuring Danny DeVito as "The Penguin."

actors were used in the chariot race scene to create the realistic crowd; today similar results can be obtained much faster and more cheaply with computer simulations of this type: large collections of autonomous interacting agents. Of course, the fact that these simulations are inspired by biological systems and appear to give a relatively good simulation of such systems is reason to consider these models as something more than a way to give Hollywood cheap and dramatic backgrounds to its movies. These simulations provide a mechanism to begin to ask more fundamental questions about the way in which groups of animals interact.

In shoals of fish (and other animals), it has long been supposed that there is some sort of "safety-in-numbers"; indeed it is reasonable to suppose that this is the reason for the existence of such large groups in the first place. However, it is not clear exactly where this benefit comes from. Certainly, if you are part of a large group, then the chance of getting swallowed by a passing shark is less — but the chance of attracting a shark in the first place is greater. There are two competing theories to explain the benefits to the shoal of having a large number of members [45]: the larger the group, the greater the chance of having a particularly astute group members; or, alternatively, simply that more eyes are better. Recent experimental evidence (in [45]) has lent weight to the second explanation. In an experimental study, the authors of [45] studied a group of fish in a tank with two paths — one leading to a fake predator, and the other not. They measured the ability of individual fish to avoid the predator p and then found that a group of n fish would avoid the predator with a probability closely matched to that predicted theoretically $1 - (1-p)^n$. But this experimental result is also a confirmation of the model of flocking we have adopted here. The collective behaviour and coherent behaviour in Figs. 9.3 and 9.4 are a result of a homogeneous ensemble of agents: each individual agent (fish or bird) behaves in exactly the same way, and yet the collective behaviour of the entire ensemble benefits exactly as predicted.

Similar insight has been gained when these flocking models are applied to human movement. Of course, the understanding of human movement has far greater relevance than academic zoological curiosity. In Chapter 8, the reaction diffusion model could be successfully applied to model phenomena such as "Mexican waves": when the crowd of a sports stadium spontaneously organises itself to periodically rise and sit, creating the effect of a wave. But, of much greater practical importance is to understand the movement of people in congested traffic flow (either pedestrian or vehicular) and during emergency evacuation.

In [13] the authors examined the modelling of highway traffic. They assumed that cars would travel in lanes subject to basic traffic rules and an individual target speed. The cars would choose to change lanes if doing so was advantageous (i.e. would allow them to travel faster). Under these very simple rules, it was shown that the traffic behaviour was basically dependent on density. As the density of cars increased, the free-flowing traffic would be replaced by coherent moving blocks of cars – this is essentially what one will typically experience in real freeway traffic. But, importantly, this also allows one to make predictions about what density, and also speed, of traffic would be achievable for a given road and traffic configuration.

Essentially the same observations have also been made for models of human traffic [24]. Groups of pedestrians have been observed and found to exhibit similar density dependent behaviour — consistent with models of the same phenomenon. While this experimental verification of these models is comforting, it is only qualitative — essentially, we are seeing that the model simulations "look right." In the next section we will start to address the second (currently, not fully answered) question of how we determine whether the model is really a good model or not.

9.3 Evaluating Crowd Simulations

In Chapter 4 we briefly described how one could build predictive models of time series data. Such models could be easily evaluated based on how well they actually worked — how well they predicted new events (see Section 4.6). Our problem is that now we cannot apply the same criteria to the models presented here because the models are not models of a single time series and it is not just the predicted value that is important.

What we want to know is how well the *behaviour* of the models replicates the behaviour of the real systems. In the previous section we argued that the predictions looked right. But, how do we quantify that? There is no, as yet, clear answer. But one reasonable approach would be to find some quantity of real interest: measure how the flock works together, how big is it, or how dense. Then, we may make qualitative comparisons to real observed data.

There is a second, more philosophical objection to the models we have proposed in this chapter. Essentially, the rules on which the models are based, and the models themselves, have just been "invented." Once we created a model we then, as a final step, compare it to the data. In Section 4.6 the models were constructed in the reverse order: we start with observed data, fit the models to the data and then make observations on the model structure. We can do the same here. However, first the positional data must be recorded from the real group of animals (or people) and this is something that is only now becoming possible (see [1, 26]).

9.4   Summary

The study of collective dynamics is still a new and rapidly evolving field, and
we have only just provided a very basic description of some of the more fundamental models. The main message of this chapter is to compare and contrast
the models we have presented here with the models in previous chapters.
These models of collective behaviour are agent based: the rules are applied
to each individual in the much larger collection. Moreover, the rules have
simply been invented to mimic reality. Nonetheless, the comparison to reality
is good; much like the reaction-diffusion systems discussed previously, these
simple models are capable of mimicking a wide variety of realistic behaviours.
In Section 9.3 we briefly described some further steps that will be necessary
to verify the model behaviours we observe here, and do correspond to the
correct models of the underlying biological process.

Glossary

Agent An individual, whose dynamics follows a set of simple rules.

Flock A collection of agents: a flock, herd, school, shoal, swarm, crowd,
plague, pride, etc.

Exercises

1. Confirm that $x(x-1)e^{-x}$ conforms to the general shape of Fig. 9.2.

2. Modify the MATLAB® code `boidswarm.m` (available from the website)
   to generate the Vicsek model of [43].

3. Computationally (or if you really want to, analytically) confirm that the
   collective behaviour observed in the simulation programme `boidswarm.m`
   does indeed depend on agent density.

4. Modify the MATLAB® code `boidswarm.m` to create a model of highway
   road traffic, as described in the text.

5. What happens to the collective behaviour of Rules 1, 2, 3 if each of these
   rules (in turn) is ignored?

6. In Fig. 9.3, different qualitative behaviours were observed for the same system (and the same parameters on the rules). Why do you think this is? Is this likely to be true for *any* choice of parameter values?

Chapter 10

Complex Networks

10.1   Human Networks: Growing Complex Networks

Complex networks are ubiquitous. In biological systems, the brain is a network of neurons connected by synapses, the cardiovascular system is a vast network of veins and arteries, the foodchain is a vast interconnected web of predators and prey, within a single cell, the complex inter-reaction of various chemical actions and signals serves as a dynamic cellular communication system; and, disease transmission between individuals follows specific infection pathways which characterise the spread of an infection.

The science of complex networks has arisen from the mathematical field of graph theory. In graph theory (and hence in complex network science), a *network* (or *graph*) is a set of *nodes* connected by *edges*. The nodes may represent individual neurons and the edges the synapses connecting them. Or, in disease modelling, the nodes are (infected) individuals and the edges are infection pathways. One of the earliest examples of such mathematical modelling with a network is due to Leonard Euler (at least according to Wikipedia[1]). In the city of Kaliningrad (which once was known as Königsberg), there are (or were, in 1735) seven bridges connecting both banks of the Pragel River and two islands. Is it possible to construct a walk through the city crossing each bridge once and only once? Euler constructed a simple network and showed that no such path existed (since more than two nodes in the network had an odd number of links). Euler's transformation of the topography of Königsberg into a network is depicted in Fig. 10.1.

Figure 10.1 illustrates the reduction of a physical problem to representation as a network, and then the further reduction of that network to a matrix encoding the connectivity information between nodes in the network. The *Adjacency Matrix* A_{ij} is defined as follows: element a_{ij} is 1 if nodes i and j are connected, and zero otherwise. Of course, in the Königsberg bridge problem there is, in some cases, more than one bridge connecting two landmasses, and so the definition in Fig. 10.1 is generalised to also encode this information

---

[1] See http://en.wikipedia.org/wiki/Seven_Bridges_of_Königsberg.

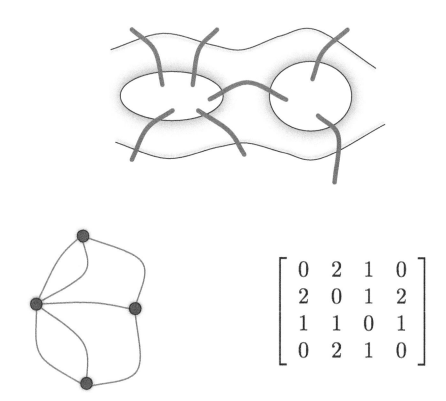

Figure 10.1: **The seven bridges of Königsberg.** The upper panel depicts the north and south bank of the river Pragel, together with the two islands and the seven bridges. In the lower panels this is reduced to a network of four nodes and seven links. The nodes correspond to the four landmasses and the links to the seven bridges. Finally, we represent this as an *adjacency matrix*: an n in row i column j indicates that landmass i is linked with landmass j by n distinct bridges, and 0 indicates that there are no bridges joining these landmasses.

about the degree of connectivity. However, when all that is important is whether or not two nodes are connected, then the binary matrix is sufficient.

While mathematicians have studied graphs such as those in Fig. 10.1, the field of complex network science, and its application to biological systems has only really developed since the start of the twenty-first century – largely as a result of the development of sufficient computational power to be able to study much larger networks. Königsberg only has seven bridges[2] connecting four distinct nodes. In the human brain there are some 10^{12} nodes and around 10^{15} links. For diseases transmitted in human communities, the number of nodes will typically exceed several million (see Sections 10.2 and 10.3). Such large networks obviously require substantial computational resources to be able to understand and model them. The most basic question is then how one characterises the type of a network when that network consists of many millions (or more) nodes. A very large adjacency matrix would be nice – but it is not always possible.

Instead, the approach which is often taken is to characterise the network into one of several basic types. Of these, four types will be of most interest to us:

- Random networks,

- Regular networks,

- Small world networks, and

- Scale-free networks.

We will deal with each in turn. Typical networks of each of these four types (genera) are illustrated in Fig. 10.2.

A *random* network consists of a set of nodes with completely random connections between them. Each node is connected to some others, but exactly which others it is connected to is determined randomly. Thus, the adjacency matrix is a binary matrix which is symmetric ($A_{ij} = A_{ji}$) but otherwise the locations of the non-zero elements is random — see Fig. 10.3. A *regular* network is exactly the opposite – it exhibits some strong regular structure. Nodes are connected to neighbours with some logical and repeating pattern. For example, the nodes in (B) are arranged in a ring (i.e. ordered from 1 to 100) and each node is connected to the two nodes on either side of it. A *small world* network can be constructed from a regular network be rewiring a small number of links (as in Fig. 10.2(C)). Finally, a *scale-free* network can be constructed by building a small random network and then adding new nodes and new links such that each new node (to be added incrementally) is more likely to be connected to high degree nodes. Adjacency matrices for each of the four

---

[2]In modern Kaliningrad, that number has actually been reduced to five, and only two of these are original.

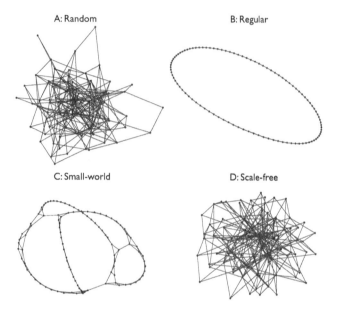

Figure 10.2: **Four different "types" of networks.** The four networks represented here (with nodes as small dots and the links as solid lines) depict small (100 node and about 500 links) networks for the case of: (A) random connection between nodes; (B) regular "lattice-like" connection (in this case, nodes are linked in a chain to the two nodes before and after them); (C) a small world network (constructed by randomly rewiring a small portion of the links in (B)); and, (D) a scale-free network. Both (A) and (D) exhibit random wiring, the difference is that in (D) the links are preferentially attached to the most highly connected nodes: this creates a greater concentration of links to a smaller number of nodes, whereas in (A) the links are more evenly distributed among the nodes — this is evident from the picture.

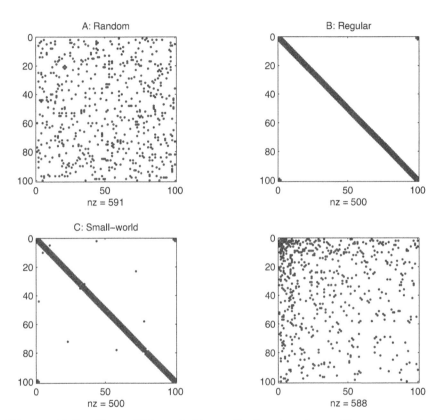

Figure 10.3: **Four different "types" of networks: adjacency matrices.** The adjacency matrices for the four networks represented in Fig. 10.2: (A) random connection between nodes, (B) regular "lattice-like" connection, (C) a small world network, and (D) a scale-free network. In each case, every element of each matrix is either 0 or 1. Non-zero elements are depicted by solid dots, zeros are not shown.

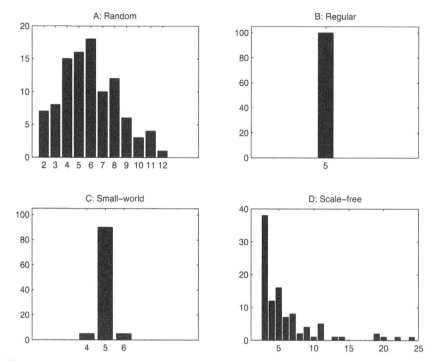

Figure 10.4: Four different "types" of networks: degree distributions. The degree distributions for the four networks represented in Fig. 10.2.

networks shown in Fig. 10.2 are illustrated in Fig. 10.3. It is important to note that we have not *yet* defined either scale-free or small world networks — we have only described one method to contruct such things, without saying what it is that we are constructing.

So that we can make a precise definition of what we mean by both "scale-free" and "small world" we first need to define several properties of complex networks. These properties are useful in characterising various features of a network — without worrying too much about details of specifically how a particular network is wired.

First, the *degree distribution* of a network characterises the distribution of links amoung nodes. The degree distribution is a probability distribution function $P(n)$ such that $P(n) = \text{Prob}(\text{a randomly chosen node has } n \text{ links})$. The degree distributions for the four networks of Fig. 10.2 are illustrated in Fig. 10.4. For the random network, the degree distribution is binomial[3]: this is because the location of the links (the 1's in the adjacency matrix)

---

[3]That is, it follows a binomial distribution: the probability of k successes from n trials is proportional to $p^k(1-p)^{n-k}$ is the probability of success in a single event is p.

are distributed randomly, and hence a given column of the adjacency matrix (the number of links a particular node has) is equivalent to the number of 1's (which occur with a fixed probability p in n trials — where n is the number of nodes). The degree distribution for the regular network is regular, and the degree distribution for the small world network is almost regular too (except for the small number of random rewirings). Finally, the scale-free distribution has a strongly skew distribution: most nodes have a small number of links but a small number have a very large number: in fact, by definition this distribution is a *power-law distribution.*

A power-law distribution is a probability distribution $P(n)$ where the

$$P(n) \propto n^{-\gamma}$$

for some constant exponent γ. Hence, if the probability distribution is plotted on a logarithmic scale ($\log P(n)$ vs. $\log n$), the relationship between the log-variables is linear

$$\log P(n) \propto -\gamma \log n.$$

A power-law distribution is also known as a *scale-free distribution*, and a network which exhibits a power-law degree distribution (for any $\gamma > 0$) is a *scale-free network*. For these reasons, when considering the possibility of a scale-free distribution of node degree, one would typically look at the log-log plot of degree — as in Figs. 10.8 and 10.14.

Why do we care about power-law distributions and scale-free networks? It is probably easy to think of real-world situations that would give rise to random and regular networks, but what about scale-free networks? In fact, power-law distributions occur in a very wide variety of settings and their defining property — most having little and a few having very much — is seen in many settings, particularly social ones. Personal income (or wealth, however you choose to measure it) follows a power-law distribution very closely: most people have a relatively small income, but a small number have a very large amount of money. In sociology and economics, this is known as the *80-20 rule* (or the *Pareto principle*, after the economist that developed these ideas): 80% of the wealth is owned by 20% of the people, or, 20% of the effort (work) produces 80% of the result. Applied to complex networks, we interpret this to mean 20% of the nodes have 80% of the links. Because this principle is applied for any particular subset of the whole, there is an invariance of scale, and hence it is "scale-free".

The really striking thing about scale-free distributions and scale-free networks is that they really do occur in a very wide range of areas. One of the reasons for this is the simple *Preferential Attachment rule* for generating such networks. Generating a scale-free network is essentially a very simple two-step process:

1. Start with a small network of m nodes, each connected to all the others (typically, $m = 4$ will do).

2. Iteratively add new nodes, one at a time. For each new node, make k connections to existing nodes in the network. The probability of connecting the new node to existing node i is proportional to $d(i)$, the degree of node i.

The *degree* of a node is the number of links a node has. Hence, nodes with more links are more likely to get even more, and nodes with few links are relatively less likely to get additional links. The phenomenon is known as "rich-get-richer", and this intuitive description explains why scale-free distributions are so prevalent in economic settings: money makes money.

At the very far tail of the scale-free distribution are the super-rich. In economics, these are individuals with incomes vastly beyond everything else. In the theoretical scale-free distribution, this means that there is a small but finite probability of encountering infinitely large values. In the scale-free network this means that there is a finite possibility of finding values infinitely large (of course, for a given finite network, there will be a particular largest value – but as the network grows, so do these largest values).

Degree distribution is sufficient to characterise scale-free networks, but not small world. Loosely speaking a small-world network is one for which most neighbours of a node are neighbours themselves (like a regular network), but at the same time, the average number of connections between two randomly chosen nodes is small (like a random network). To quantify these two features we define the *average path length* (and the related *diameter*) and the *clustering coefficient* (and the related but distinct *assortativity*) for a random network. These four computational quantities allow us to characterise a network, and, in particular, we expect a small-world network to have a low average path length and high clustering coefficient[4].

Path length is the distance between two nodes in the network. So, average path length is the average distance between randomly chosen pairs of nodes in the network. Choose two nodes in the networks (at random) and then ask how many links does one have to traverse to get from one to the other. The diameter of a network is the largest such distance between two points in the network. Figure 10.5 illustrates the concept of path length. From Fig. 10.5 we can write down a table indicating connectivity between pairs of nodes — Table 10.1. This table can then be directly transposed to provide the following adjacency matrix (where, by convention, we place 1s on the main diagonal —

---

[4]Although the current scientific literature is still vague on how small is "low" and how big is "large."

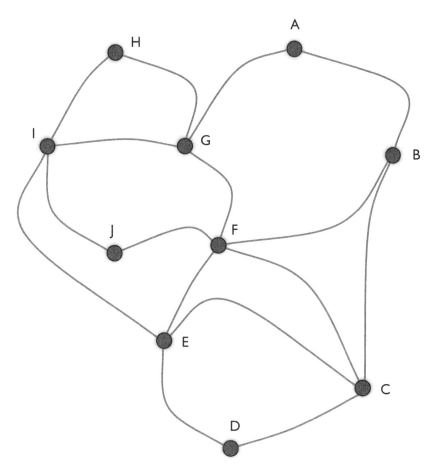

Figure 10.5: **Path length and diameter.** The shortest path between nodes A and C is 2 (via node B). The diameter of the network is 3 as this is the largest *direct* distance between any two nodes (for example, A and D).

TABLE 10.1:   **Adjacencies for the nodes of the network of Fig. 10.5.**

	A	B	C	D	E	F	G	H	I	J
A	1	1	0	0	0	0	1	0	0	0
B	1	1	1	0	0	1	0	0	0	0
C	0	1	1	1	1	1	0	0	0	0
D	0	0	1	1	1	0	0	0	0	0
E	0	0	1	1	1	1	0	0	1	0
F	0	1	1	0	1	1	1	0	0	1
G	1	0	0	0	0	1	1	1	1	0
H	0	0	0	0	0	0	1	1	1	0
I	0	0	0	0	1	0	1	1	1	1
J	0	0	0	0	0	1	0	0	1	1

indicating that node N_j is accessible from node N_i):

$$
A = \begin{bmatrix}
1 & 1 & 0 & 0 & 0 & 0 & 1 & 0 & 0 & 0 \\
1 & 1 & 1 & 0 & 0 & 1 & 0 & 0 & 0 & 0 \\
0 & 1 & 1 & 1 & 1 & 1 & 0 & 0 & 0 & 0 \\
0 & 0 & 1 & 1 & 1 & 0 & 0 & 0 & 0 & 0 \\
0 & 0 & 1 & 1 & 1 & 1 & 0 & 0 & 1 & 0 \\
0 & 1 & 1 & 0 & 1 & 1 & 1 & 0 & 0 & 1 \\
1 & 0 & 0 & 0 & 0 & 1 & 1 & 1 & 1 & 0 \\
0 & 0 & 0 & 0 & 0 & 0 & 1 & 1 & 1 & 0 \\
0 & 0 & 0 & 0 & 1 & 0 & 1 & 1 & 1 & 1 \\
0 & 0 & 0 & 0 & 0 & 1 & 0 & 0 & 1 & 1
\end{bmatrix}.
$$

For such a small network it is easy to perform an exhaustive computation of the pairwise pathlength between all nodes. If the (i, j)-th element of A (which we will denote as (A_{ij})) is non-zero, this indicates that node-i can be reached from node-j in one step. Similarly, if the (i, j)-th element of $A^2 = A \times A$ is non-zero, then this indicates that node-i can be reached from node-j in two, or fewer, steps. In this example all elements of A^3 are non-zero (but this is not true for A^2), and hence the network has diameter 3. Of course, for larger networks where exhaustive computation via matrix multiplication is infeasible, these quantities can be estimated by randomly sampling pairs of nodes.

Whereas path length and diameter measure the "size" (or perhaps "spread" would be a better term) of a network, assortativity and clustering coefficient measure the local structure of a network: these quantities measure how much the neighbours of a node know about one another.

The clustering coefficient is defined[5] to be the probability that two randomly

---

[5]Actually, there are several variants of this definition, but, for the sake of simplicity, we adopt the most direct.

chosen neighbours of a node are neighbours themselves. That is,

$$C = \text{Prob(neighbours of node } k \text{ are neighbours)}$$
$$= \text{Prob}(A_{ij} = 1 | A_{ik} = 1 \text{ and } A_{jk} = 1)$$

$$= \frac{1}{N} \sum_k \left(\frac{1}{\sum_{\ell \neq k} A_{\ell k}} \sum_{\substack{i, j \\ A_{ik} = A_{jk} = 1}} \frac{A_{jk}}{\sum_{\substack{i, j \\ A_{ik} = A_{jk} = 1}} 1} \right).$$

The first two expressions are equivalent. The second expression, albeit un-wieldy, simply shows that this can be computed directly by counting elements A_{ij} of the adjacency matrix A (A is an $N \times N$ matrix). Of course, the clustering coefficient can also be computed by counting the number of triangles in the network (since neighbours who are neighbours will form a triangle). The number of such triangles compared to the maximum possible gives the clustering coefficient.

Assortativity measures the tendency of nodes to be connected to other nodes of similar degree. Unlike clustering, which is measured based purely on counting structural properties of the network, assortativity is a measure of the degree to which statistical properties are preserved between neighbours. There are many different statistical properties of the nodes which could be measured in this way, but the only one which is currently of interest to us (and the one which is most commonly meant when talking about assortativity) is the node degree. For all nodes i and j such that they are neighbours, plot $d(i)$ against $d(j)$. The correlation coefficient of that scatter plot is defined to be the assortativity of the corresponding network. In Fig. 10.6 we illustrate such plots for the four networks of Fig. 10.2.

Table 10.2 summarises the sample values for the various statistics we have discussed for each of the four networks of Fig. 10.2. As expected, path-length and clustering tend to go hand-in-hand. If a network is highly clustered (like the regular network (B)), then the average path-length is quite large. Conversely, networks that are completely random (A) will tend to have small average path-lengths. Scale-free and small world networks represent special cases — the scale-free network will have even-lower-than-random path-length as the large degree nodes act as hubs and allow for even more efficient movement through the network. The small world network combines a high clustering coefficient (representing a strongly regular structure) and a *relatively* small path-length[6].

Small average path-length and strong clustering define small world networks. But the origin of these networks[7] is due to some early work by so-

---

[6]This effect is more noticeable in larger small world networks. The example we have chosen here has only 100 nodes and so the effect is not very pronounced.

[7]And, the as, yet poorly explained title of this section.

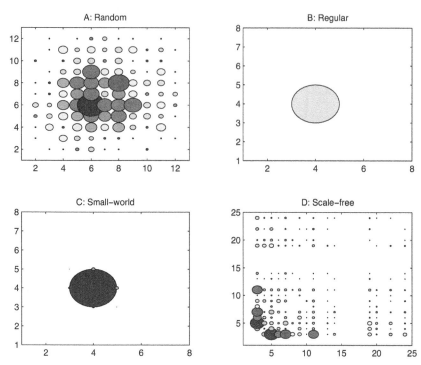

Figure 10.6: Degree-degree correlation and assortativity. Plots of
the frequency of occurence of pairs of nodes sharing a common link with the
indicated degrees. For node-i and node-j (such that $A_{ij} = 1$), the horizontal
and vertical axes are the corresponding node degree $d(i)$ and $d(j)$. The radius
of the circle on the plot is proportional to the frequency with which nodes of
degree $d(i)$ and $d(j)$ are linked. For the regular and small world networks,
there is a very strong uniformity in network degree (see Fig. 10.4) and so the
information conveyed by this diagram is minimal. For random and scale-free
networks, there is slightly more information here. For the random network
we see correlation between node degrees; for scale-free networks we see some
anti-correlation (as the large degree nodes are bound to connect to the more
prevalent low-degree nodes). The assortativities computed for three of these
four networks are (A) 0.0666, (C) -0.0566, (D) -0.0496. Because the nodes'
degrees of (B) have zero variance, correlation is not well defined; moreover, the
significance of the measured value for (C) is also questionable. Nonetheless,
the slight negative correlation is typical of preferential attachment networks,
and the slight positive correlation is to be expected for a random network.
See colour insert.

TABLE 10.2: Summary of statistics computed from the networks of Fig. 10.2.

	A	B	C	D
average path-length	2.79	12.75	7.60	2.73
diameter	7	26	16	6
clustering coefficient	0.0712	0.5	0.459	0.1033
assortativity	0.0666	*	-0.0566	-0.0496

ciologists looking at connections between individuals in a society[23][8]. The sociologist Stanley Milgram observed that the social connections between people are actually very strongly interdependent. Through a rather ingenious[9] experiment, he gave an envelope to a random group of people and told them that it should be sent to a specific person — whom none of them knew. Each of the individuals was then asked to forward this envelope to one of their own acquaintances whom they thought would be more likely to know the intended target. This process was then repeated with the new recipients until the envelope did, eventually, reach the actual target. Milgram found that, across the whole of North America (with a population of around 250 million), the average number of steps required to achieve this was only six.

This fact has been picked up in popular culture and cited, or mis-cited, on many occasions. The simplest version: any two people are connected through a chain of only six acquaintances — there are "six degrees of separation" between any two people on earth. Nonetheless, this is exactly the property that has led to the development of small world networks: most of your friends are friends themselves (high clustering) and yet there is a short chain of acquaintances connecting you to everyone else[10] (short average path-length). Nonetheless, this property remains true to many interconnected systems and there are several popular examples of it: notably, the Kevin Bacon game (connections among Hollywood actors) and the Erdös number (connections between mathematicians).

With the Kevin Bacon game, Hollywood actors are linked if they co-star in the same movie. The objective of the game is to link a nominated actor to Kevin Bacon through a chain of movies in which pairs of them co-star. The Erdös number follows the same principle and aims to link mathematicians if they co-author a paper. The objective is to find the minimal number of papers linking an individual to Paul Erdös, a particularly prolific mathematician with work spanning a diverse range of areas. The number of connections required

---

[8]Several statements of this idea were made prior to Milgram, but these tended to be essentially literary rather than scientific in nature. Interestingly, the idea of a scale-free network can also be traced back much further than the current interest in such things, to some early papers published in the field of bibliometrics [6, 7].

[9]If not methodologically flawed.

[10]That is, almost everyone else. There are certainly counter examples and exceptions.

to link an individual with Erdös is known as the Erdös number. The Erdös number of Erdös himself is 0, his 511 co-authors have an Erdös number of 1, their co-authors have Erdös number of 2. It is estimated that 90% of the world's active mathematicians have an Erdös number of 8 or less. My Erdös number is 3.

In the next section we move on from the harmless, but pointless, pastimes of mathematicians and show how ideas from network science can be applied to model the spread of disease. In Section 10.2 we examine an abstract model of the spread of infection (SARS) on a small world network and argue that it closely matches reality. In Section 10.3 we examine data from avian influenza and show that it matches what one would expect for a scale-free network.

10.2   Small-World Networks and the Spread of SARS

The SEIR model discussed in Section 7.3 supposes that all individuals have an equally small probability of being infected. In this section, we propose an alternative model structure where infection can only occur along certain predefined paths. Suppose that the population is arranged in a regular grid. Each node can infect its four immediate neighbours (horizontally and vertically) and a random number of remote nodes; this is a small-world network like that in Fig. 10.2(C), except that here we use a regular two-dimensional grid rather than a one-dimensional chain.

We suppose that the probability of infecting near neighbours (supposed to be members of the same household) is likely to be distinct from the probability of infecting remote neighbours (supposed to be daily acquaintances). Moreover, it is clear that by adjusting the relative magnitude of these two probabilities, one can generate transmission dominated by "clusters" (i.e. localised transmission).

Let p_1 denote the probability of infecting each of the near neighbours of a node and let p_2 denote the probability of infecting the remote neighbours. Suppose that there are n_1 near neighbours and n_2 remote neighbours. Moreover, we consider the model with four distinct groups of individuals, S, E, I and R, corresponding, as before, to those that are susceptible, exposed, infected and removed. This situation is depicted in Fig. 10.7. The probability of transitioning from state E to state I on a given day is r_0, and the probability of transitioning from I to R is r_1.

We allow n_2 to be random and fixed for each node, hence node-i has $n_2^{(i)}$ links. Moreover, to explicitly model the small-world structure with a scale-free network:

$$P(n_2^{(i)} = e^k) = \frac{1}{\mu} e^{-\frac{k}{\mu}}. \tag{10.1}$$

(a) State transition flow graph

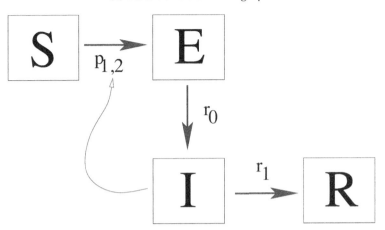

(b) Small world network structure

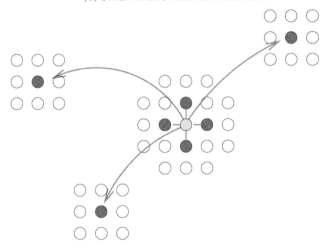

Figure 10.7: The four compartmental small-world network model of disease propagation. The top panel depicts the transmission state diagram: S to E based on the SW structure and the infection probabilities $p_{1,2}$; E to I with probability r_0; and, I to R with probability r_1. The lower panel depicts the distinction between short-range and long-range network links. The lower panel shows the arrangement of nodes in a small network. The central (light grey) infected node may infect its four immediate dark grey neighbours with probability p_1 and three other red nodes with probability p_2. (Figure reproduced with permission from [38].)

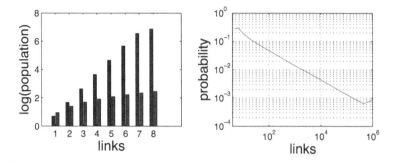

Figure 10.8: Typical small world and scale-free distributions. The left panel depicts the computational estimate of the number of nodes connected to a given node by n or fewer links (taller bars). Also shown is the number of nodes connected to a given node by n or fewer links if we allow only local transmission along any of the neighbouring paths (proportional to $(2n - 1)^2$). The right panel depicts the distribution (estimated from a simulation of 10^6 nodes) of the number of links that a given node has. Parameter values are $n_1 = 4$, $\mu = 2.4$ and $n = 2700$. (Figure reproduced with permission from [38].)

Hence, although the near neighbour links are bi-directional, the remote neighbour links are only one-way.[11] In Fig. 10.8, we depict the induced small-world network structure and the explicitly modelled scale-free structure of the network. Namely, the number of links between any two nodes in the model is small; and the probability that a node has a given (large) number of links is exponential. From Fig. 10.8(a), we see that 85% of nodes are connected by less than 8 links (and "almost all" are connected by no more than 8 links). In fact, we choose the number of neighbourhood links so that the connectivity of the social network in Hong Kong would be roughly the same as the "six-degrees of separation" observed in North America [23].

In the simulations that follow we always take $n_1 = 4$ to indicate the immediate horizontal and vertical neighbours and $N = 2700$ so that $N^2 \approx 7.3 \times 10^6$ is approximately the population of Hong Kong. Variation of the parameter r_0 is largely immaterial and we therefore fix $r_0 = \frac{1}{6.4}$. With a daily probability of transition from E to I of $\frac{1}{6.4}$, the average incubation period is 6.4 days (as reported by [8]).[12] The remaining parameters of the model p_1, μ, p_2, r_1 we study in more detail in the next section.

---

[11]Furthermore, for computational convenience they are only assigned for nodes that become infected.

[12]Admittedly, the 95% confidence interval is 1–18 days and is therefore significantly greater than that reported by [8]. However, we found that the model simulations were remarkably robust to changes in r_0.

We fix $\mu = 2.4$ as this seems to be roughly appropriate for the degree of social connectivity observed in human networks. As a first approximation, we also set $p_1 = 0.02$ as suggested by [18]. We do not adopt $p_2 = 0.37$ as cited by [48] as this case is probably somewhat atypical. Moreover, this value leads to rapidly explosive growth of the epidemic, which is highly unlike the observed behaviour. According to published reports [8], the time before hospitalisation is 3 to 5 days. If we assume that hospitalisation is equivalent to isolation (and therefore transition from I to R), we can take $r_1 = \frac{1}{4} = 0.25$.

Denote by $\hat{n}_2$ the expected value of n_2. From the distribution in Eqn. 10.1, for $\mu = 2.4$, this is approximately 8. The parameters n_1, p_1, $\hat{n}_2$, p_2 and r_1 can be used to approximate the expected number of new infections $E(-\Delta S)$ by

$$E(-\Delta S) = (n_1 k p_2 + \hat{n}_2 p_2 - r_1)I, \qquad (10.2)$$

where k is the number of near neighbour links that support possible infections and, because the near neighbour infections are arranged in "clumps", $\frac{1}{2} \leq k \leq 1$. Moreover, for each infected node, the number of new secondary infections per day is approximately $(n_1 k p_2 + \hat{n}_2 p_2)$ and the total will be

$$n_k r_1 + 2n_k(1 - r_1)r_1 + 3n_k(1 - r_2)^2 r_1 + \ldots = \frac{n_k}{r_1},$$

where $n_k = (n_1 k p_2 + \hat{n}_2 p_2)$. From the available data [30], the average number of secondary infections is 2.7; therefore, we take

$$n_1 k p_2 + \hat{n}_2 p_2 = 2.7 r_1,$$

and hence $p_2 \approx 0.386 r_1 - 0.08$. For $3 < \frac{1}{r_1} < 5$ this gives $0 < p_2 < 0.05$.

By studying the stability of the difference equation version of Eqns. (7.20) through (7.23), we obtain the eigenvalues

$$\lambda_1 = 1,$$

$$\lambda_{2,3} = 1 - \frac{r_0 + r_1}{2} \pm \sqrt{\frac{1}{4}(r_0 - r_1)^2 + n_k r_0} \qquad (10.3)$$

where, as before, $n_k = n_1 k p_1 + \hat{n}_2 p_2$ is the average number of infections. We note that, as expected, the disease will be contained if $n_k < r_1$. Suppose that the (average of) 2.7 infections occurred prior to hospitalisation after an average of 4 days, then $n_k = 2.7/4 = 0.675$. Hence the rate of spread of infection is given by

$$1 - \frac{r_0 + r_1}{2} + \sqrt{\frac{1}{4}(r_0 - r_1)^2 + 0.675 r_0}.$$

Figure 10.9 compares the average growth rate observed in the data (a 5-day moving average of the ratio of the total number of infections on two successive days) to that computed from these eigenvalues. From the first 40

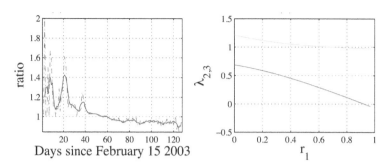

Figure 10.9: Rate of growth of SARS in the data and eigenvalues of the model. The left panel is an estimate of the rate of growth in the number of infections from the data in Fig. 7.6. The instantaneous estimate is shown as a dashed line; the 5-day moving average is shown as a solid line. The right panel gives the eigenvalues of Eqn. (10.3). Growth rates, comparable to that observed in the data (i.e. 1.11 to 1.2), are observed for $0.2 < r_1 < 0.025$.

days of data for Hong Kong, the mean rate of infection is 1.19 and the range is approximately 1.1 to 1.42. Rates of infection of 1.19 and 1.1 correspond to r_1 values of approximately 0.05 and 0.2, respectively. Hence, we conclude that this model indicates that infection did not cease with hospital admission after between 3 and 5 days. In the early stage of the epidemic, the rate of transmission of the virus indicates that patients remained infectious for much longer periods of time (possibly up to 20 days).

We now examine the role of clustering in the spread of infection. Clearly, the number of "clumps" that occur in the model will be proportional to the number of long-range infections. As such, we expect the ratio of clusters to infections is approximately given by $\frac{\hat{n}_2 p_2}{n_1 k p_1 + \hat{n}_2 p_2}$. If $\hat{n}_2 p_2 \approx 0$, the infection is largely localised and the growth of infection is polynomial (see Fig. 10.8). If $\hat{n}_2 p_2 \ll n_1 k p_1$, the growth of infection will still eventually become exponential (Fig. 10.8) but initially the spread is largely local and polynomial. Finally, if $\hat{n}_2 p_2 \gg n_1 k p_1$, the rate of growth is exponential and the spread of infection is equivalent to a stochastic version of the standard SEIR model. It is the intermediate dynamics for $\hat{n}_2 p_2 \approx n_1 k p_1$ that are of most interest to us. In Fig. 10.10, we illustrate the typical spread of infection for each of these four scenarios. From Fig. 10.10 one observes that increasing $\hat{n}_2 p_2$ increases the number of clumps and also the rate of transmission of infected individuals. With $p_1 = 0$ we see only isolated infections with no spatial correlation, and, conversely, for small $\hat{n}_2 p_2$ we see a small number of large clumps. In all simulations we see that the rate of infection is approximately polynomial until several nonlocal infections occur. At this point, one sees an explosive (exponential) growth in infection.

We now present simulations of our model with various parameter values and

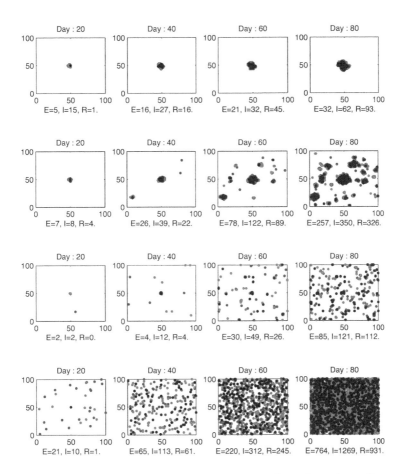

Figure 10.10: **Simulations of disease spread under the effect of a small world.** Each row depicts the evolution (at 20-day intervals) of infected individuals for different parameter values. In all cases, $N = 100$ (to ease visualisation), $n_1 = 4$, $\mu = 2.4$, $r_0 = \frac{1}{6.4}$ and $r_1 = 0.05$. Infected nodes are shown in dark grey, exposed nodes in light grey and removed nodes in black. The top row is with $p_1 = 0.2$ and $p_2 = 0$, the second row is with $p_1 = 0.2$ and $p_2 = 0.002$, the third row is with $p_1 = 0.02$ and $p_2 = 0.02$, and the fourth row is with $p_1 = 0$ and $p_2 = 0.05$. (Figure reproduced with permission from [38].) See colour insert.

attempt to reproduce the observed dynamics. Following this, we provide some Monte Carlo simulations (i.e. random computer realisations) with the same parameter values and show that the range of observed behaviour is extremely wide.

Clearly, the original data represents a non-stationary system. One can observe at least two distinct phases. In our linear modelling (Fig. 8.1), we assumed three phases, hence for our small-world network model we also assume three phases. The reason for restricting our interest to such a small number of (discrete) regimes is to avoid problems associated with over-determined systems.[13] We could seek more realism by increasing the range of parameter values during the epidemic, but this would only be appropriate for a completely deterministic model. Instead, we take $N = 2700$, $n_1 = 4$ and $\mu = 2.4$ as before. The parameters $p_{1,2}$ and $r_{0,1}$ are set as follows:

$$p_1 = 0.02,$$

$$p_2 = \begin{cases} 0.1 & 1 \leq t \leq 30 \\ 0.02 & 30 < t \leq 60 \\ 0.01 & t > 60 \end{cases},$$

$$r_0 = \frac{1}{6.4}, \text{ and}$$

$$r_1 = \begin{cases} 0.025 & 1 \leq t \leq 20 \\ 0.1 & 20 \leq t \leq 40 \\ 0.333 & t > 40 \end{cases}.$$

The three changes of p_2 correspond to the initial phase when SARS spreads freely,[14] an intermediate phase when $p_2 = p_1$, and a final phase when $p_2 \approx 0$. In this third and final stage, a combination of limited movement of the population and rudimentary hygiene measures combine to effectively limit transmission. Similarly, we include two changes in the value for r_1. During the initial phase, r_1 is close to zero, which corresponds to the rate of growth when the disease spreads without any control measures. During the second phase, rudimentary control is in place and r_1 has decreased somewhat (as observed in Fig. 10.9). However, it is only during the final phase that $r_1 \approx \frac{1}{3}$, corresponding to removal of infected individuals approximately 3 days after becoming infectious (i.e. in this final phase, hospitalisation is equivalent to isolation). Figure 10.11 presents the result of 1000 simulations.

Finally, we provide some quantitative evidence that the simulations and model behaviour are similar. In Section 7.4 we showed that the ordinary SIR and SEIR type models exhibited dynamics unlike the observed data. In Fig. 10.11 we see that, according to the very gross measure of total infections, the small-world model and simulations are similar.

---

[13] Even for deterministic models, one needs to avoid excessive parameterisation.
[14] One could also argue that this phase also models the (initially) highly contagious individuals observed in the early stages of the outbreak. We make no such distinction.

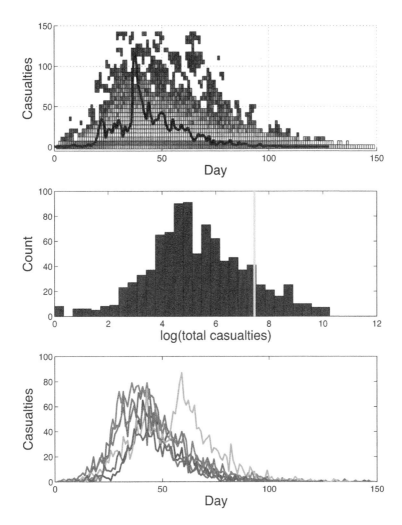

Figure 10.11: Variation in model behaviour, compared to the data.
The top panel is a probability density plot of the temporal evolution of 1000
simulations (high probability is shown as a dark band near to x-axis) compared
to the data. The middle plot is a comparison of the total number of casualties
for each of those simulations compared to the true data value (1735). Approximately 13.5% of all simulations exhibited a greater total casualty count
than the true data. The bottom plot is seven "representative" simulations
(these simulations were randomly chosen from among the 91 simulations with
a total casualty of between 1000 and 2500). In each case, we see that the data
is typical of the observed simulations. (Figure with permission reproduced
from [38].)

This SW model of disease propagation, when applied to the SAR-CoV propagation dynamics for Hong Kong in 2003, shows that the most serious effect of SARS is nosocomial transmission. By isolating infectious individuals as soon as infection is identified (typically in 3 to 5 days), the severity of the SARS outbreak in Hong Kong in 2003 would have been significantly lessened.

Moreover, from a theoretical viewpoint, we see that the crucial feature in capturing the dynamic behaviour of the transmission dynamic time series is combining local and non-local transmission. We find that although super-spreader events, where one individual infects a large number of others, are typically associated with highly infectious individuals, they can be modelled equally well with highly *connected* individuals. It has been widely reported that highly connected individuals will occur in human society and this may be a more appropriate way to model bursty propagation of diseases within communities.

The model presented in this section provides a complex network-based counterpoint to the homogeneous models discussed previously (in Chapter 7). However this model, like those in Chapter 7, is an invention which was subsequently shown to do a good job of modelling certain real situations. In the next section we do the reverse: we start with data from the real situation, and, from that, derive an appropriate model.

10.3   Global Spread of Avian Influenza

Standard mathematical models of geographical transmission of an infectious agent assume that the terrain is locally homogeneous and that the pathogen will diffuse uniformly [25]. A natural consequence of this formulation is that if the transmissibility of the pathogen is lower than some threshold, the disease will terminate. Recent studies of infectious agents (usually either biological pathogens or computer viruses) in certain complex networks have shown that in these networks such a threshold does not exist. In particular, if the connectivity within a network follows a scale-free distribution and the transmissibility of the agent is positive, then an epidemic is inevitable [2, 22]; we will discuss this again in Section 10.4. For the case when recovery from the infected state confers immunity, an epidemic is inevitable only if the population is infinite [22] or if the system is not closed — which, since the life cycle of domestic poultry is relatively short, is the situation for avian influenza. More-over, one can expect that the infection will persist indefinitely. Whereas, if a disease is spreading according to standard mathematical models with uniform mixing, then it can be eradicated by lowering the rate at which it is trans-mitted below a predetermined, but non-zero, threshold. However, if a disease is spreading on a scale-free network, the eradication of that disease is only possible if transmission is reduced to precisely zero.

In this section we consider the network consisting of individual communities (the network nodes) explicitly connected by possible infection pathways (the links). We then focus only on the communities which have become infected, but study all possible links between them. Our analysis of data from the geographical and temporal distribution of avian influenza outbreaks exhibits a network that is scale-free. That is, the number of links $k \geq 1$ has probability distribution

$$P(k) = \frac{k^{-\gamma}}{\zeta(\gamma)}$$

with $\gamma > 1$. The denominator $\zeta(\gamma)$ is Riemann's zeta function[15] and provides the appropriate normalisation constant. Note that if $1 < \gamma \leq 2$, this distribution does not have a finite mean. Even if $2 < \gamma \leq 3$, the variance of the number of links is infinite and therefore even with very small (but non-zero) rate of transmission, transmission will still persist [22].

When considering disease transmission, we treat the nodes on the network as susceptible individuals or communities and the links between them are potential transmission pathways. Two nodes are linked if transmission between them is possible. Ideally, we should treat the actual transmission pathways. But, that information is not available. We therefore assume that transmission can occur only over a local area (in both time and space). We consider both the un-weighted network (where all transmission pathways are considered as "possible") and a weighted version (where we consider the plausible pathways and weight according to the number of incoming connections to a node). To address the limitations of our data we also test our method against a short time span late in the outbreak (to eliminate inconsistency between earlier reports of outbreaks) and large outbreaks (to minimise the effect of aggregation of reports). In all cases the results are both qualitatively and quantitatively similar.

To the best of our knowledge, there is no data that confirms that any disease can be transmitted in this way. However, there is significant evidence that the communities which support various infections do exhibit scale-free structures. Experimental evidence shows that human travel (and therefore human contacts) exhibits scale-free structure [3] and that networks of sexual contacts (number of sexual partners) is also scale-free [19]. Computer simulations have shown that simulation in this manner is viable, and suggested potential containment strategies [10]. In this section, we focus on transmission of the avian influenza virus in wild and domestic bird populations. There is good reason to suppose that transmission of this virus between bird flocks may follow a

---

[15]The Riemann zeta function $\zeta(s)$ is defined for complex s such that

$$\zeta(s) = \sum_{n=1}^{\infty} \frac{1}{n^s}.$$

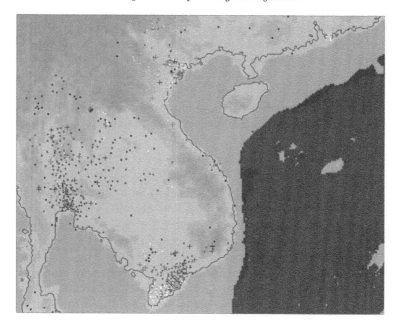

Figure 10.12: Avian influenza case data. Part of the data used in this study, overlayed against a crude map of the coastline of East Asia. Human cases are marked with solid dots, animal cases with crosses. Colour coding is by date. The three large clusters correspond to the outbreaks in Cambodia and in the north and south of Vietnam (around Hanoi and Ho Chi Minh City), respectively. Hainan island is marked in the north-east of the image and outbreaks in Hong Kong are visible as a light grey spot in the far north-east corner. (Figure reproduced with permission from [40].)

scale-free distribution [16]. Nonetheless, it is currently not obvious that the traditional and alternative uniform-mixing models are inadequate.

The data we use in this study are a compilation of all reported avian cases of avian influenza between 25 November 2003 and 10 March 2007. The data consists of 3346 recorded cases. For each case, the date of the outbreak and the location (longitude and latitude) are recorded. Individual cases may either be wild birds that are found (possibly post-mortem) and determined to be infected with a strain of avian influenza, or the detection of an avian influenza strain in a domestic flock (most probably then followed by culling of that flock). Data relating to the magnitude of each incident are also recorded. Human cases of avian influenza have also been recorded in the same data set, but for this study, these are ignored. The entire data set is compiled from a variety of sources[16]: Figure 10.12 depicts one snapshot.

The data consists of 3346 triples of the form (t_n, λ_n, ϕ_n), where t_n is the time (in days) since 25 November 2003 of the n-th incident. The parameters λ_n and ϕ_n are the latitude and longitude of that case. Each incident (t_n, λ_n, ϕ_n) corresponds to a node on the graph of infection links. We construct a directed link from node-i (t_i, λ_i, ϕ_i) to node-j (t_j, λ_j, ϕ_j) if

$$d(i,j) \le (t_j - t_i)\mu \qquad (10.4)$$

and

$$0 \le (t_j - t_i) < T_{\max}, \qquad (10.5)$$

where $d(i,j)$ is the great circle distance between node-i and node-j in kilometres and μ is a positive constant (units of km/day) corresponding to the approximate geographical rate of transmission of the virus. Great circle distance is computed from longitude and latitude using standard spherical geometry,

$$d(i,j) = R \times \dots$$
$$\arctan\left\{ \frac{\sqrt{[\cos\phi_j \sin\Delta\lambda]^2 + [\cos\phi_i \sin\phi_j - \sin\phi_i \cos\phi_j \cos\Delta\lambda]^2}}{\sin\phi_i \sin\phi_j + \cos\phi_i \cos\phi_j \cos\Delta\lambda} \right\},$$

where $\Delta\lambda = \lambda_j - \lambda_i$ and the radius $R = 6372.795$ km.

The choice of the criterion (10.4) and (10.5) to determine connectivity is, of course, arbitrary. But, it is also natural. If we assume that the geographical rate of transmission of the virus is uniform and equal to μ, then node-i is deemed to be connected to node-j if the virus at node-i can travel as far as node-j before the outbreak is observed to occur at node-j (that is, within $t_j - t_i$ days) and sooner than $T_{\max}$ days. We have varied both parameters μ and $T_{\max}$ over a wide range of values ($3 < \mu < 50$ km/day and $5 < T_{\max} < 30$ days) and have not found significant qualitative variation in the results. We take $T_{\max} = 10$ days and $\mu = 25$ km/day. The choice of 25 km/day is motivated by the apparent rate of spread of avian influenza cases in the early stage of the outbreak. The choice of 10 days is only to provide more easily visible results. Larger values only make the network denser; smaller values make it more fragmented.

With these values of $T_{\max}$ and μ we construct a complex network of connectivity of avian influenza cases. The connectivity between individual nodes in that network is shown in Fig. 10.13. In Fig. 10.13(a), the diagonal structure of the matrix indicates connection between temporally adjacent nodes. The clustering in Fig. 10.13(b), is due to geographical localisation. The *sample*

---

[16]The data originally come from World Organisation for Animal Health alerts (see http://www.oie.int/) and World Health Organisation case reports and are all manually entered using ArcGIS and converted to Keyhole Markup Language (KML) using Arc2Earth (http://www.arc2earth.com/). The data are available, in a format compatible with Google Earth (KML), from http://www.declanbutler.info/Flumaps1/avianflu.html.

(a) date–ordered   (b) unordered

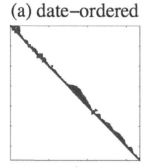   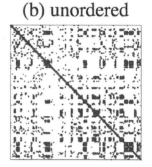

Figure 10.13: **Network connectivity (adjacency) matrix.** Both panels depict the connections present in the network deduced according to the criteria (10.4) and (10.5). If there exists a connection from node-i (vertical axis) to node-j (horizontal), then the point (j, i) is marked. In panel (**a**), the nodes are ordered according to time (that is, $i < j$ only if $t_i \leq t_j$). In panel (**b**), the points are ordered approximately geographically. (Source: Figure reproduced with permission from [40].)

average number of connections from a given node is (a relatively large) 16.8 and because of our criterion for selecting connectivity, nodes are connected only if they are separated by no more than 10 days. Hence, the fact that the data span 1203 days indicates that the shortest path between random nodes can be very large: hence, this is not a small world network. However, the reason for this is entirely artificial. The geographical connectivity may well be small world (as illustrated for humans in [3] and various livestock populations in [5, 17]), but because we constrain nodes in time, this feature is suppressed. However, the available data make it impossible to resolve this issue.

Nonetheless, the resultant network is scale-free. This is evident from Fig. 10.14. Figure 10.14(a) illustrates that this network is composed of discrete clusters. The two main reasons for this disconnectedness is our initial assumptions concerning connectivity (10.4) and (10.5) and the inevitable incompleteness of available data.

In Fig. 10.14(b) we depict the link distribution and an estimate of the scale exponent. Following [11] we estimate the exponent γ using a maximum likelihood estimator which avoids statistical bias associated with a linear fit to the log-log plot[17]. By altering $T_{\max}$ or μ, we can change γ, but changing

---

[17]In [11], the authors show that the maximum likelihood best value of γ is the solution to

$$\frac{\zeta'((\gamma)}{\zeta(\gamma)} = -\frac{1}{N}\sum_{i=1}^{N} \log(x_i),$$

where x_i is the number of links associated with the i-th node.

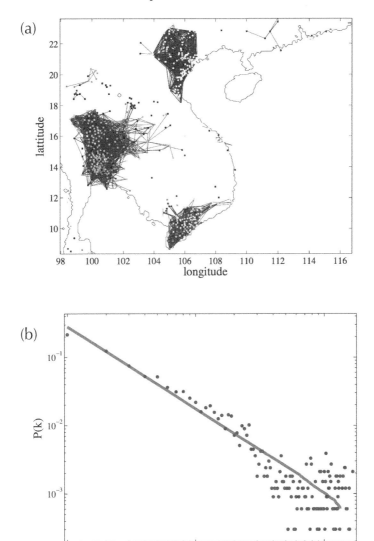

Figure 10.14: **Network degree distribution.** In panel (a), the data from Figure 10.12 are re-drawn with the addition of network connections. Clearly, the entire network is not connected. Nonetheless, from this network we compute the degree distribution (panel (b)) and display it on log-log scale. The data exhibit a scale-free distribution with estimated scale exponent of $\gamma \approx 1.2028$. The Kolmogorov–Smirnov (KS) goodness-of-fit test indicates a value within the 90% confidence interval given that the underlying data is sampled from a power-law distribution. (Figure reproduced with permission from [40].)

these parameters does not affect our main result: the network is scale-free and has infinite mean and variance. Conversely, increasing this average number of connections or choosing a more complicated metric (rather than Eqn. (10.3)) can increase the connectedness of the final network.

To examine the robustness of our result, we also examined the networks obtained only from the most recent data (18 months, since October 2005) and from large outbreaks (more than 50 deaths). These tests resulted in a restriction of our dataset to 1942 and 899 nodes respectively. In both cases, similar scale exponents (~ 1.2) and KS-test confidence levels were obtained. Finally, we also considered the construction of a weighted network. Each node was weighted by the reciprocal number of incoming links, and the degree of each node was set to be the sum of these weights. The degree distribution we obtained in this way was also approximately scale-free ($\gamma \approx 2.51316$), but because of the high degree of clustering in the network the relationship only extended over one decade.[18]

It is worth noting that the distance between outbreaks, with $d(i,j)$, $d(i,j) + \mu(t_i - t_j)$, or $|d(i,j) - \mu(t_i - t_j)|$ as a norm, is not scale-free: it is multi-modal and decays approximately exponentially. Hence, the scale-free features we observe are not due to the spatial (or temporal) distribution of outbreaks, but rather to the large variability in infectious pathways in the complex network topology. We also observe a fairly low scale exponent $\gamma \approx 1.2$. This is lower than the oft-cited "typical" range of $2 < \gamma \leq 3$, but of the same order of magnitude as experimental results for human travel [3] ($\gamma \approx 1.6$). This exponent is also similar to the network scaling ($\gamma \approx 1.8$) reported for e-mail collaborative networks [9].

Note that we are not able to trace the actual infection pathways. Instead, we take the observed data for outbreaks of avian influenza and construct a network that *contains* part of the underlying transmission paths. We assume that the virus propagates at a constant and relatively modest rate, and related events must be relatively close in time. Certainly, delays in detection and reporting of cases, and long-distance transmission (for example, via migratory birds) would violate these assumptions. Hence the network we construct is inevitably only an approximation, and the inclusion of these additional factors could result in a more connected and more realistic network structure. Nonetheless, to do this would introduce many more parameters and cloud the basic result: the spatial-temporal connectedness (defined by Eqns. (10.4) and (10.5)) is scale-free.

---

[18]We have also used this data to construct a fully connected graph, first by eliminating singleton nodes, and then by connecting discrete clusters (closest first).The resulting graph is still scale-free ($\gamma \approx +1.32$) and exhibits strong clustering and high assortativity.

10.4   Complex Disease Transmission and Immunisation

Given that a disease is spreading on a complex network (either scale-free, or small world or both), how does this affect our response to that disease? How should immunisation be managed to exploit the complex network structure most effectively? In which sorts of network structures *can* infection be controlled? How does this differ from the homogeneous approach to disease modelling introduced in Section 7?

Scale-free networks have nodes with very high degree: there is a small, but finite probability of encountering nodes with arbitrary large degree. If we apply this idea to the modelling of disease transmission, there is a finite possibility of encountering individuals with a very large number of transmission pathways to others, and therefore a very high likelihood of spreading any contagion.

Let us consider a scale-free network; the probability of a randomly chosen node having n links is given by $kn^{-\gamma}$ where the constant k is such that the probabilities sum to one $k = \sum_{n=1}^{\infty} n^{-\gamma}$ (that is, $k = \zeta(k)$, the Reimann zeta function). Then, what is the average degree of a randomly chosen node? The mean degree μ is given by

$$\mu = k \sum_{1}^{\infty} n \times n^{-\gamma} \qquad (10.6)$$

$$= k \sum_{1}^{\infty} n^{1-\gamma} \qquad (10.7)$$

and the variance is

$$\sigma^2 = k \sum_{1}^{\infty} (n - \mu)^2 \times n^{-\gamma}. \qquad (10.8)$$

So, when $\gamma < 3$, the variance is not finite, when $\gamma < 2$ neither is the mean. When considering the transmission of a disease over such a network, this is extremely important. Whereas, in Section 7.3 we saw that SIS dynamics will give rise to a threshold and transmission below that threshold will disappear; for scale-free networks (with $\gamma < 3$) this threshold is always 0 [2]. Hence, such diseases will always persist and can only be eliminated by destroying the scale-free structure of the network.

However, this is something of a simplification. Whether a disease is uncontrollable in a real situation is somewhat more complicated than the physicists' model of an infinite complex network. First, any real disease is only being transmitted over a finite network. Hence, the threshold, while small, will still be non-zero. Second, while scale-free networks are widely found in nature, they are probably not ubiquitous. There are many examples of physical systems which appear to have a scale-free (power-law) degree distribution, but for which that degree distribution may actually either be exponential (it is difficult to tell the two apart from data), or power law, only up to some certain maximum. In social networks, for example, while a small number of individuals will have a very large number of links, these numbers can always be bounded. No matter how friendly one is, it is only possible to have a finite number of friends in a finite life span.

Thirdly, the presence of scale-free-ness in the connectivity of individuals within the community may alter the response of both individuals and the community to disease. Suppose (as was done in [51]) that vaccination is also available and is completely voluntary. Moreover, suppose that each individual in the community will choose to vaccinate or not, based on their personal perceived risk. Then, if we construct a risk function which is proportional to node degree — since one with more connections (such as a primary school teacher) has a higher risk than an individual with fewer connections (such as an author) — then quite naturally high degree nodes become more likely to vaccinate. If the rate of vaccination increases with node degree, the effect of scale-free networks in perpetuating disease spread [2] is completely reversed [51]. Nodes with higher degree actually become immunised from the network and the disease propagates more slowly, or becomes eradicated. In effect, the scale-free distribution loses its long tail as an effect of voluntary vaccination.

So far in the discussion of complex networks, we have focussed on disease transmission, but we have also taken pains to emphasise the near ubiquity of network science and the wide range of possible applications. In the remainder of this chapter we briefly mention some of those applications: complex networks in music, sheep and neural systems.

10.5   Complex Networks Constructed from Musical Composition

Music, of course, is not a biological system and as such falls outside the scope of this book. But, our subjective response to music is most certainly the result of the dynamic response of a particular biological entity to the dynamic variation in a tonal sequence. Moreover, music is a product of the ultimate biological system – human society.

Consider the sequence of notes in a musical composition. Each note is defined by its pitch (frequency) and its duration (crotchet, quaver, semi-quaver and so on). Let these notes correspond to nodes on a complex network. Nodes are connected by a link if the corresponding notes occur successively in the

TABLE 10.3: **Properties of networks constructed from musical compositions.**

work	mean degree	mean shortest path	diameter	clustering coefficient	node degree distribution exponent
Bach (Violin solos)	14.7	2.9	10	0.35	1.4
Bach (WTC I)	20.1	2.9	8	0.26	1.3
Bach (WTC II)	16.8	2.8	9	0.48	1.4
Chopin (Op. 23)	17.4	3.0	12	0.36	1.0
Chopin (Nocturne)	15.9	3.5	15	0.15	1.4
Mozart (Assorted)	15.1	3.3	13	0.24	1.3
Jay Chou ("Secret")	10.6	3.7	19	0.24	1.4
Teresa Teng ("Hits")	9.8	3.8	16	0.17	1.8

Adapted, with permission, from [20].

Figure 10.15: **Musical composition from a complex network.** Using Bach's violin solos (No. BWV1002) to construct a network, an artificial score is then composed by randomly resampling nodes on the network following the connectivity pattern of the network. (Figure reproduced with permission from [20].)

Day 15   Day 21

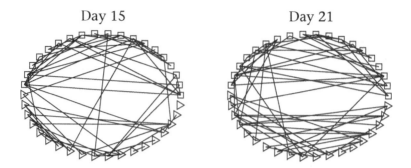

Figure 10.16:   **Interaction patterns for grazing sheep.** Male sheep are
denoted by triangles and females by squares. Links indicate an interaction –
that is, the two animals are doing the same thing. This work [44] indicates
much stronger intra-sex rather than inter-sex interaction among the animals.
(Figure reproduced with permission from [44].)

composition. With this simple rule we can construct a complex network from
a musical score or set of scores. There are two interesting consequences of
this [20]. First, across musical genres, and cultural norms, for music which
is considered appealing, the corresponding complex networks exhibit similar
qualitative features. From Bach and Chopin to Chinese pop: the networks
exhibit a power-law exponent in node degree; and a range of network statistics (including assortativity, clustering and degree) each fall within the same
range — see Table 10.3. Second, with this network inferred from a work (or
more readily a body of work attributed to a single composer). It is possible to construct new musical compositions in a particular style by randomly
traversing the network of musical notes. Compositions constructed in this
way can[19] be subjectively viewed as appealing, and also be judged typical of
the particular artist or composer from which the sample was originally drawn.
Such a reconstructed composition is illustrated in Fig. 10.15.

10.6   Interaction of Grazing Herbivores

In this chapter we have extensively examined the interaction of animals or humans under the effect of disease transmission processes. In the previous Chapter we looked at how animals can (perhaps) interact during natural movement.

---

[19]Although with some caveats, see [20].

Here we examine the grazing behaviour of herbivores: a herd of sheep, eating [44]. By measuring the grazing behaviour of each sheep it is possible to infer connections between them. Here, we define grazing behaviour by a binary variable: head up or head down, indicating that either the animal is eating or not. This gives us a binary times series for each individual. We then measure interaction between animals by looking for patterns of similar behaviour. This leads to a network of interactions between specific sheep: Figure 10.16. From this computation it is possible to observe sex-dependent interactions among sheep. Male-male interaction and female-female interaction is stronger than interaction between males and females. Importantly (for ecologists), this behaviour is not dependent on the size of the animals (competing hypotheses would suggest that animals of a similar size would act similarly).

10.7   Neuronal Networks Are Complex Networks

This chapter could not be complete without briefly mentioning the most obvious example of a complex network, particularly in the context of this volume: the brain. Brain networks cross multiple scales; from individual connections between neurons to function networks of brain areas — and can be classified as either anatomical or functional. Functional brain networks can further be classified as those founded on correlation (just like the sheep in Section 10.6 individual areas are deemed linked if their behaviour is similar) or causation (like the musical network, nodes are linked if there is a causative link between them[20]).

Brain networks can be deduced by detailed, anatomical studies of regions of brain tissue, to list definitive links between distinct neurons. Clearly this is a huge task, but it is a task which has been completed once, for one organism: *Caenorhabditis elegans* or *C. elegans*, the famous microscopic transparent roundworm which is widely used by biologists and neuroscientists alike as a model organism. It is a favoured model organism precisely because it is so simple (well, relative to "higher" organisms, that is). *C. elegans* has around 1000 cells in its entire body, the fully developed adult is known to have precisely 302 neurons, and the wiring diagram for those 302 neurons is now known precisely[21]. Of course, with a precise wiring diagram, as depicted in Figure 10.17, there is hope that computational simulations can then be produced to simulate the creature. However, even for this simple worm[22], this is not

---

[20]In analysing brain networks, such a causative link is actually often based on some sort of statistical correlation and so the delineation between these two areas is blurred.

[21]See http://wormweb.org/neuralnet for the full anatomical network of this nemotode worm's "brain."

[22]With a relatively limited social and intellectual life.

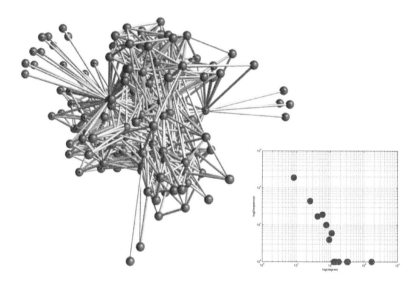

Figure 10.17: **Neural network of *C. elegans*.** We plot here the full neural network (unweighted, but derived from the weighted network of [46, 47]); and the logarithmic distribution of node degree (showing an approximately linear distribution of node degree on the double logarithmic scale).

possible.

10.8   Summary

In this chapter we have barely touched the surface of the rapidly growing area of complex network science. Section 10.1 introduced some basic concepts, some models for building complex networks, and some useful statistical measures. In Section 10.2 and 10.3 we embarked on a detailed analysis of two specific epidemics and how they can be usefully modelled using complex networks. In Section 10.2 the model was empirical and shown to closely match the data observed for SARS in Hong Kong during 2003. In Section 10.3 the model was motivated by data and fitted to that data. This analysis allowed us to show that, in fact, the distribution of bird flu outbreaks follows a power-law distribution. The remaining sections of this chapter diverged from the study of diseases and presented applications of complex network science to other areas: music, animal behaviour and neuroscience. Each of these applications opens up a vast range of potential applications (see, for example, [42] for an overview of complex network methods to neuroscience). This chapter can hardly do justice to all of these areas.

Glossary

Assortativity Assortativity is a measure of to what extent connected nodes share similar values of some attribute. Most often, the attribute of interest is node degree. In this case, assortatitivy can be defined as the statistical correlation coefficient between node degree of linked nodes.

Average path-length The average minimum number of links required to connect a randomly chosen pair of nodes in a network.

Clustering coefficient The probability that two nodes which are both connected to a third will also be connected to one another.

Edge The connections between those objects are the edges.

Network A mathematical graph, a collection of nodes and edges. A network provides a mathematical description for the interconnection among a set of objects.

Network diameter The maximum, over all pairs of nodes, of the minimum number of links required to connect any pair of nodes in a network.

Node The individual objects of a network are represented as nodes.

Pareto principle To paraphrase, the Pareto principle states that most of the effect can be attributed to a small minority. Also referred to as the 80-20 rule, in which case the proportions can be quantified: 80% of the effect is due to 20% of the actions.

Power-law A power-law distribution is one where the probability of observing a value n is proportional to $n^{-\gamma}$ for some (positive) constant γ. Power-law distributions are characterised by a small but finite probability of arbitrary large events.

Preferential attachment The principle under which a scale-free network will emerge naturally. New nodes should connect to existing nodes in a network with a preference for attaching to nodes with higher degree.

Random network A random network is one in which the edges are arranged randomly between pairs of nodes.

Regular network A regular network, such as a grid, consists of nodes connected in an orderly and consistent fashion. Typically this will mean that the nodes are arranged in a chain (in one dimension), grid (in two dimensions) or array (in higher dimension).

Scale-free network A scale-free network is defined to be a network such that the degree distribution follows a power law.

Small-world network A small-world network is one which is almost regular, but with the addition (or substitution) of a few random links. The defining characteristics of a small-world network is a small average path length (similar to that which would be expected for a random network) and strong clustering (similar to what one would expect for a regular network).

Exercises

1. If the cadiovascular system were to be described as a network, to what do the nodes and edges correspond?

2. Compute the mean and variance for a scale-free distribution with exponent α.

3. Compute assortativity and clustering coefficient for the network example of Fig. 10.5.

4. Compute average path-length as a function of network size for preferential attachment networks. What do you observe?

5. Give reasons why the popular idea of "six degrees of separation" cannot be inferred directly from the work of Stanley Milgram.

6. Confirm, analytically, that the sums in Eqn. (10.6) and Eqn. (10.8) do not converge for $\gamma < 2$ and $\gamma < 3$ respectively.

Chapter 11

Conclusion

11.1 Models Are a Reflection of Reality

But, models are not reality. We began this text by discussing different types of models, and these models can be used to help us understand reality. Yet, at every step, when we build a model, we make a simplification of the thing being modelled. Each model is an approximation. In the words of George Box: "All models are wrong, some models are useful." It is not that the models lie, merely that they are economical with the truth, and the art in appropriate application of models is in finding the best economies. The model should be simple enough to be useful, but not so simple that it no longer reflects the interesting parts of reality.

In modelling neuronal dynamics (Chapter 6), for example, we seek models that are capable of expressing dynamical behaviour consistent with a single neuron. We do not (yet) have the knowledge or technical tools necessary to build a model of human consciousness. Yet if we wish to learn about the function of individual isolated neurons, then that level of detail is unnecessary.

Conversely, the models of disease transmission introduced in Chapter 7 are perfectly adequate for some diseases, but not others. They model influenza epidemics very well, but are unable to explain the data observed for SARS in Hong Kong. More complex models, using complex networks, are able to explain many of the features of transmission of SARS (Chapter 10). But these models too, are economical with the truth. In 2009 a new strain of swine flu (H1N1 influenza) spread from Mexico and raised global concern. In June of 2009, the World Health Organisation declared it a pandemic. Fear, confusion and panic spread as health experts and computational models predicted a dangerous and severely infectious global wave of infection. The models were wrong. While the H1N1 strain of influenza did go on to infect millions of people globally, and led to deaths in excess of 15,000, this should be compared to a background of several hundred thousand deaths from all other forms of influenza. But, in this case, the model is only wrong because the parameters (that is, the data which drive the model) was not accurate. For the parameter values which were employed, the behaviour of the model was entirely accurate, and, in the future a similar infectious outbreak could still illicit the response predicted by the model.

The problem with models then is rather a problem of obtaining sufficient data of high enough quality. Science is now reaching a stage were the increasing availability of data is beginning to match the ability of modelling methods such as these to explain the underlying phenomena. In Chapter 9 we saw examples of models of collective behaviour. It is now possible to monitor (via global positioning satellites, for example) individual birds in a flock and thereby build realistic models of such behaviour — not just at the level of qualitative similarity as has been done in the past.

New genetic information is increasingly becoming available — from the study of the human genome and the genomes of many other model organisms. The genetic basis for life is now within reach of modelling methods. The Physiome project[1] seeks to describe, at the same level of detail, physiological cellular processes and provide models of them. In Chapter 5.1 we described some very basic cellular level chemistry. The work of the Physiome project goes far beyond this. These are computational models of every aspect of cellular life. As our understanding of the data and the biology improves, then our models become better liars.

In Fig. 10.17, for example, we are able to show the wiring diagram for the "brain" of a nematode worm. Of course, we are still a long way from constructing wiring diagrams for the human brain, but we are getting better. Nonetheless, even as we approach that goal, the models will still remain only as models. Just as with simpler organisms (Fig. 10.17), these simple wiring diagrams are not enough to fully explain behaviour. In those situations where biological understanding remains inadequate, we are forced to fall back on building our models from data.

Building models from data is a two-part problem. The first part of the problem is obtaining the data itself. In Chapter 3 we reviewed some basic electrophysiological phenomena that allow one to monitor the changing dynamical state of different systems within the human body. In terms of mathematics, there is some hidden set of differential (or difference) equations driving the dynamical behaviour which we observe. That system of equations could be very large, but it remains essentially hidden from view. Instead, we are limited by our ability to only measure the output of that system of equations — maybe only just one variable which we are able to observe. In Chapter 4 we showed that, with just one variable measured over a sufficiently long time, it is possible to access the hidden equations.

By measuring one variable over time, we are able to infer not just the value of that variable but also its derivatives: how fast it is changing, how fast the change in that variable is changing and so on. If we do this sufficiently many times (obtaining sufficient number of derivatives) and then keep watching the system for long enough, we will finally obtain a single sample trajectory from the underlying "hidden" system. From that single trajectory we can build

---

[1]See http://www.physiome.org/.

a computational model (Section 4.6). Provided that trajectory is sufficiently long, the computational model will be a reflection of the "hidden" rules of the underlying system. Computationally, the model we have obtained is described by Eqn. (4.24) and therefore is not quite as pretty as an elegant set of differential equations would be. But it will do the same job. The model will lie well enough to reproduce the expected behaviour, and then we can learn something about the system which the model describes.

Of course, this model can only work well provided that what it is expected to lie about does not change. That is, the model can be expected to *interpolate* but not *extrapolate*. The model will continue to make good predictions of the future, provided that the future continues to look like the past. Once the future ceases to look like the past, then the model will be reduced to guessing at what is an appropriate behaviour. In Chapter 2 we discussed this problem. We looked at how the system behaviour can often depend on a particular bifurcation parameter. For some particular value of that bifurcation parameter, the model can work well. But, if you change the bifurcation parameter, then the system behaviour can change, potentially quite dramatically, and then all bets are off [39].

Bibliography

[1] M. Ballerini, N. Cabibbo, R. Candelier, A. Cavagna, E. Cisbani, I. Giardina, A. Orlando, G. Parisi, A. Procaccini, M. Viale, and V. Zdravkovic. Empirical investigation of starling flocks: A benchmark study in collective animal behaviour. *Animal Behaviour*, 76:201–215, 2008.

[2] M. Boguna, R. Pastor-Satorra, and A. Vespignani. Absence of epidemic threshold in scale-free networks with degree correlations. *Physical Review Letters*, 90:028701, 2003.

[3] D. Brockmann, L. Hufnagel, and T. Geisel. The scaling laws of human travel. *Nature*, 439:462–465, 2006.

[4] M. Chan-Yeung and R.-H. Xu. SARS: Epidemiology. *Respirology*, 8:S9–S14, 2003.

[5] R. M. Christley and N. P. French. Small-world topology of UK racing: The potential for rapid spread of infectious agents. *Equine Veterinary Journal*, 35:586–589, 2003.

[6] D. J. de Solla Price. Networks of scientific papers. *Science*, 149:510–515, 1965.

[7] D. J. de Solla Price. A general theory of bibliometric and other cumulative advantage processes. *Journal of the American Society for Information Science*, 27:292–306, 1976.

[8] C. A. Donnelly et al. Epidemiological determinants of spread of causal agent of severe acute respiratory syndrome in Hong Kong. *Lancet*, 361:1761–1766, May 7 2003.

[9] H. Ebel, L.-I. Mielsch, and S. Bornholdt. Scale-free topology of e-mail networks. *Physical Review E*, 66:035103, 2002.

[10] S. Eubank, H. Guclu, V. V. A. Kumar, M. V. Marathe, A. Srinivasan, Z. Toroczakai, and N. Wang. Modelling disease outbreaks in realistic urban social networks. *Nature*, 429:180–184, 2004.

[11] M. Goldstein, S. Morris, and G. Yen. Problems with fitting the power law distribution. *European Physics Journal B*, 41:255–258, 2004.

[12] W. Gurney, S. Blythe, and R. Nisbet. Nicholson's blowflies revisited. *Nature*, 287:17, 1980.

[13] D. Helbing and B. A. Huberman. Coherent moving states in highway traffic. *Nature*, 396:738–740, 1998.

[14] Y. Hirata, N. Bruchovsky, and K. Aihara. Development of a mathematical model that predicts the outcome of hormone therapy for prostate cancer. *Journal of Theoretical Biology*, 264:517–527, 2010.

[15] K. Judd and A. Mees. On selecting models for nonlinear time series. *Physica D*, 82:426–444, 1995.

[16] W. B. Karesh, R. A. Cook, E. L. Bennett, and J. Newcomb. Wildlife trade and global disease emergence. *Emerging Infectious Diseases*, 11:1000–1002, 2005.

[17] I. Z. Kiss, D. M. Green, and R. R. Kao. The network of sheep movements within Great Britain: Network properties and their implications for infectious disease spread. *Journal of the Royal Society Interface*, 3:669–677, 2006.

[18] J. T. Lau et al. Probable secondary infections in households of SARS patients in Hong Kong. *Emerging Infectious Diseases*, 10:235–243, 2004.

[19] F. Liljeros, C. R. Edling, L. A. N. Amaral, H. E. Stanley, and Y. Åberg. The web of human sexual contacts. *Nature*, 411:907–908, 2001.

[20] X. Liu, C. K. Tse, and M. Small. Complex network structure of musical compositions: Algorithmic generation of appealing music. *Physica A*, 389:540–548, 2010.

[21] R. M. May. Simple mathematical models with very complicated dynamics. *Nature*, 261(459), 1976.

[22] R. M. May and A. L. Loyd. Infection dynamics on scale-free networks. *Physical Review E*, 64:066112, 2001.

[23] S. Milgram. The small world problem. *Psychology Today*, 2:60–67, 1967.

[24] M. Moussaïd, N. Perozo, S. Gairner, D. Helbing, and G. Theraulaz. The walking behaviour of pedestrian social groups and its impact on crowd dynamics. *PLoS ONE*, 5:e10047, 2010.

[25] J. D. Murray. *Mathematical Biology*, volume 19 of *Biomathematics Texts*. Springer, 2nd edition, 1993.

[26] M. Nagy, Z. Ákos, D. Biro, and T. Vicsek. Hierarchical group dynamics in pigeon flocks. *Nature*, 464:890–894, 2010.

[27] R. Penrose. *Shadows of the mind: A Search for the missing science of consciousness*. Vintage, Hopkinton, MA, 1995.

[28] D. C. Reddy. *Biomedical Signal Processing: Principles and Techniques*. McGraw-Hill, New York, 2005.

[29] C. W. Reynolds. Flocks, herds, and schools: A distributed behavioral model. In M. C. Stone, editor, *Computer Graphics*, volume 21, pages 25–34. SIGGRAPH'87, 1987.

[30] S. Riley et al. Transmission dynamics of the etiological agent of SARS in Hong Kong: Impact of public health interventions. *Science*, 300:1961–1966, 2003.

[31] M. Sarovar, A. Ishizaki, G. R. Fleming, and K. B. Whaley. Quantum entaglement in photosynthetic light-harvesting complexes. *Nature Physics*, 6:462–467, 2010.

[32] L. Sigler. *Fibonacci's Liber Abaci, Leonardo Pisano's Book of Calculations*. Springer, Berlin, 2002.

[33] M. Small. Nonlinear dynamics in infant respiration. Ph.D. Thesis, University of Western Australia, Department of Mathematics, 1998. URL: http://small.eie.polyu.edu.hk.

[34] M. Small. *Applied Nonlinear Time Series Analysis: Applications in Physics, Physiology and Finance*, volume 52 of *Nonlinear Science Series A*. World Scientific, Singapore, 2005.

[35] M. Small and K. Judd. Comparison of new nonlinear modelling techniques with applications to infant respiration. *Physica D*, 117:283–298, 1998.

[36] M. Small, K. Judd, M. Lowe, and S. Stick. Is breathing in infants chaotic? Dimension estimates for respiratory patterns during quiet sleep. *Journal of Applied Physiology*, 86:359–376, 1999.

[37] M. Small, H. Robinson, I. Kleppe, and C. K. Tse. Uncovering bifurcation patterns in cortical synapses. *Journal of Mathematical Biololgy*, 61:501–526, 2010.

[38] M. Small and C. K. Tse. Small world and scale free model for transmission of SARS. *International Journal of Bifurcation and Chaos*, 15:1745–1755, 2005.

[39] M. Small and C. K. Tse. Feasible implementation of a prediction algorithm for the game of roulette. In *Asia-Pacific Conference on Circuits and Systems*. IEEE, 2008.

[40] M. Small, D. M. Walker, and C. K. Tse. Scale free distribution of avian influenza outbreaks. *Physical Review Letters*, 2007.

[41] M. Small, D. Yu, N. Grubb, J. Simonotto, K. Fox, and R. G. Harrison. Automatic identification and recording of cardiac arrhythmia. *Computers in Cardiology*, 27:355–358, 2000.

[42] O. Sporns. *Networks of the Brain*. MIT Press, Cambridge, MA, 2011.

[43] T. Vicsek, A. Czurkó, E. Ben-Jacob, I. Cohen, and O. Shochet. Novel type of phase transition in a system of self-driven particles. *Physical Review Letters*, 75:1226–1229, 1995.

[44] D. M. Walker, C. Carmeli, F. Pérez-Barbería, M. Small, and E. Pérez-Fernández. Inferring networks from multivariate symbolic time series to unravel behavioural interactions among animals. *Animal Behaviour*, 79:351–359, 2010.

[45] A. J. Ward, J. E. Herbert-Read, D. J. Sumpter, and J. Krause. Fast and accurate decisions through collective vigilance in fish shoals. *Proceedings of the National Academy of Sciences USA*, 108:2312–2315, 2011.

[46] D. J. Watts and S. H. Strogatz. Collective dynamics of 'small-world' networks. *Nature*, 393:440–442, 1998.

[47] J. White, E. Southgate, J. Thompson, and S. Brenner. The structure of the nervous-system of the nematode *caenorhabditis elegans*. *Philosophical Transactions of the Royal Society of London B*, 314:1–340, 1986.

[48] T.-W. Wong et al. Cluster of SARS among medical students exposed to single patient, Hong Kong. *Emerging Infectious Diseases*, 10:269–276, 2004.

[49] World Health Organisation. Consensus document on the epidemiology of SARS. Technical report, World Health Organisation, 17 October 2003. http://www.who.int/csr/sars/en/WHOconsensus.pdf.

[50] D. Yu, M. Small, R. G. Harrison, C. Robertson, G. Clegg, M. Holzer, and F. Sterz. Complexity measurements for analysis and diagnosis of early ventricular fibrillation. *Computers in Cardiology*, 26:21–24, 1999.

[51] H. Zhang, J. Zhang, C. Zhou, M. Small, and B.-H. Wang. Hub nodes inhibit the outbreak of epidemic under voluntary vaccination. *New Journal of Physics*, 12:023015, 2010.

[52] J. Zhang, X. Luo, and M. Small. Detecting chaos in pseudoperiodic time series without embedding. *Physical Review E*, 73:016216, 2006.

[53] J. Zhang and M. Small. Complex network from pseudoperiodic time series: Topology versus dynamics. *Physical Review Letters*, 96:238701, 2006.

Index

For Product Safety Concerns and Information please contact our EU
representative GPSR@taylorandfrancis.com Taylor & Francis Verlag GmbH,
Kaufingerstraße 24, 80331 München, Germany

Printed and bound by CPI Group (UK) Ltd, Croydon, CR0 4YY
01/05/2025
01858515-0002